PLANNING IN CRISIS?

Planning in Crisis?
Theoretical Orientations for Architecture and Planning

WALTER L. SCHÖNWANDT
University of Stuttgart, Germany

Routledge
Taylor & Francis Group

LONDON AND NEW YORK

First published 2008 by Ashgate Publishing

2 Park Square, Milton Park, Abingdon, Oxon OX14 4RN
711 Third Avenue, New York, NY 10017, USA

Routledge is an imprint of the Taylor & Francis Group, an informa business

First issued in paperback 2016

British Library Cataloguing in Publication Data
Schönwandt, Walter L.
 Planning in Crisis? theoretical orientations for
 architecture and planning
 1. Planning 2. Planning - Study and teaching 3. Social
 planning
 I. Title
 361.6'1

Library of Congress Cataloging-in-Publication Data
Schönwandt, Walter L.
 Planning in crisis? : theoretical orientations for architecture and planning / by Walter L. Schönwandt.
   p. cm.
 Includes bibliographical references and index.
 ISBN 978-0-7546-7276-0
 1. Planning. 2. Planning--Study and teaching. 3. Social planning. I. Title.

 H97.S35 2007
 658.4'01--dc22

                                                          2007025284

ISBN 13: 978-0-7546-7276-0 (hbk)
ISBN 13: 978-1-138-25993-5 (pbk)

Contents

List of Figures and Tables

Preface

Everyone plans. Every building owner, every developer, every municipality, every city, every country, and every company must plan its own actions so as to allocate scarce resources—e.g., money, space, etc.—most advantageously. Ultimately, every kind of investment presupposes at least a minimum of planning.

While we thus plan day-in day-out, a formidable gulf has opened between theory and practice. Efforts to accompany and support planning scientifically have often come to a standstill at crucial points, and have remained in a state of stasis for the past decades.

This book centers on two crucial points connected to the topic of "constructs." What are constructs? This term essentially comprises concepts, propositions, contexts and theories. Constructs are the bearers of our knowledge and simultaneously the conceptual core of a planning assignment, but most of all they guide our actions in planning. They offer us knowledge and give us our bearings.

The following theses underlie this text:

First: concerning constructs for the *description* of planning (constructs *of* planning), there are currently no theoretical approaches that would allow us to efficiently interpret as many of the aspects that play a role in planning as possible. Such approaches are useful because they help us to avoid overlooking important aspects in planning.

Second: with regard to constructs for processing planning assignments (constructs *in* planning) we are often faced with the following difficulties: although constructs figure as the conceptual core of planning assignments, and therefore have a decisive influence both on the description of the initial situation in planning as well as on the solutions thereby taken into consideration, there nonetheless exists barely any work that can offer assistance and guidance in the treatment of constructs.

Against the backdrop of these two theses, this book is divided into two sections that differ in terms of presentation and degree of detail.

In order to make the first of these theses comprehensible even to those readers who have not, or have only marginally, delved into the relevant literature, the first part of this book—under the heading "Constructs For the Description of Planning"— will briefly adumbrate those planning models which have significantly governed discussions in the field of planning theory over the past decades. This overview will, at least in a cursory manner, reflect the range of planning-theoretic questions. Each of the described planning models delivered a significant impetus in a specific historical context, and are even today by no means obsolete.

This presentation however also illustrates that—since turning away from the "rational" model—discussions in planning theory have often been marked by the fact that they stress, and thereby push to the foreground, only single aspects of the planning process. These discussions have thus far not coalesced into an approach that systematically connects and efficiently integrates as many of the aspects that play a role in planning as possible.

For this reason, a systemic "Planning Theory of the Third Generation"—one that adequately reflects the complexity of the planning process to as great an extent as possible—will be introduced in broad stokes. It can essentially be traced back to Claus Heidemann and Jakob von Uexküll. This theory encompasses the entire content, which is to say the spatial, social, political, ecological and economic aspects, of the respective planning task. In addition, it takes into account the restrictions placed on our perceptual abilities and cognitive capacities, as well as the limits placed on possible intervention through planning. Beyond that, it establishes a connection to the relevant theoretical background, especially to the semiotic, epistemological and ethical components of planning. The presentation concentrates only on the main features of the theory, and the explanation of individual components consequently remains cursory and incomplete; only one or two points are generally raised concerning the individual aspects of the theory, since each part of this theory provides enough material for several books.

Against the backdrop of the second of the above described theses, the second part of this book—under the heading "Constructs for the Treatment of Planning Tasks"—takes up, and describes in detail, a specific topic in this "Planning Theory of the Third Generation": planning generally involves the attempt to, abstractly put, improve upon some state of affairs deemed wanting through the application of carefully thought-out action. This, however, presupposes that the planner has as accurate as possible an understanding of the state of affairs in question. Adequate constructs for the description of the planning problem are therefore required. After all, planning is by no means always about the use of "ready-made" constructs; rather, the required knowledge must often first be worked up or at least adapted to the respective planning situation. However, while there is plenty of work that shows, for example, how assessment processes, communicative processes etc. can be structured and organized, there is to date a lack of work that could offer assistance in developing constructs for describing the planning task and the starting-point of our planning suggestions and actions. In the second part of this book, a conceptual tool will therefore be presented which corresponds to the structure of the semiotic triangle and is capable of supporting and guiding this work, so that our speech, thoughts, and actions accord with each other to the greatest possible degree.

At this point, a remark concerning the so-called "scientific materialism" of Mario Bunge is in order. This particular approach is, especially in the second part of this book, the foundation of argumentation—albeit under a, for our purposes, simplified form. As a result of his intense engagement with this approach, many of the propositions which this author today thinks, speaks or writes can of course be traced back to Mario Bunge, without the author, however, being in a position to always specify the exact source of these propositions. Naturally, the author is himself responsible for any mistakes that arise, through simplifications or otherwise.

The author would like to thank Mario Bunge for his inspiring and extremely helpful work.

Many thanks are also due to Lukas Rieppel and Michael Rieppel, who worked so diligently to translate this volume in a prose style that retains the subtlety of the original without thereby sacrificing either clarity or readability. They, in turn, would also like to thank Mario Bunge for having recommended them for the project in the first place.

Finally, my colleagues at the Institute for the Foundations of Planning at the University of Stuttgart deserve thanks: Jens-Peter Grunau, Wolfgang Jung, Joachim Kieferle, Klaus Korpiun, Manfred Josef Pauli, Sabine Müller-Herbers, Sylvia Stieler, Katrin Voermanek and Peter Wasel. They each read and critically, as well as constructively, commented on portions of this book.

PART I
CONSTRUCTS FOR THE
DESCRIPTION OF PLANNING

Chapter 1

Seven Models of Planning[1]

Introduction

Various models of planning have dominated the discussion of planning theory in the past four decades.[2] We will briefly summarize seven of the most significant models in the first section of this book. With one exception, they are all normative, political models of planning.

Specifically, we will introduce each of the following in turn:

- The rational model of planning.
- The advocacy model of planning.
- The (neo)Marxist model of planning.
- The model of equity planning.
- The model of social learning and communicative action.
- The radical model of planning.
- The liberalistic model of planning.[3]

All of these models were developed, each one more or less an outgrowth of its predecessors, within a time-span that covers approximately four decades. It is imperative to note that all of the models listed above still find use in the practical work of contemporary planners, even if not always within the borders of a single nation-state.

The Rational Model of Planning

The rational model of planning is the source and inspiration for most of the other models, which are either a modification of the rational model or a reaction to (or against) it. For example, in their classic paper "Politics, Planning, and the Public Interest" (1955) Meyerson and Banfield describe the essential steps of this model as follows (see also Rittel 1972, 391):

---

1   In this section, "Seven Models of Planning", we will use the term "model of planning" in the sense of "comprehension, or understanding of planning."

2   To a large extent, the text of this first section follows the publications of Alexander 1992; Flyvbjerg 1998; Hall 1988; Hudson 1979; Mandelbaum, Mazza, Burchell 1996; Muller 1992; Foulton 1991a, b; Sandercock 1998; Sorensen, Day 1981 as well as Sorensen 1983.

3   This section will not cover technical models of planning such as we find in, e.g., urban planning, traffic or landscape planning etc.; nor those interdisciplinary (i.e. systems-theoretic) approaches that attempt to bridge the gap between individual specialized sub-disciplines.

1. Analyze the situation;
2. establish goals;
3. formulate possible courses of action to achieve those goals;
4. compare and evaluate the consequences of these actions.

Subsequent to the publication of Mayerson and Banfield's paper, countless variations on their model were presented and discussed among planning theorists. Various authors have designated the individual steps in different ways—sometimes subdividing them more coarsely, other times more finely—but the basic principles always remained the same (see for example Simon 1965 or Muller 1992).

The planning process need not always be carried out in the precise order enumerated above. In addition, each individual step includes iterative loops as well as a way of braking it up into several more manageable sub-tasks. Furthermore, various different alternatives for treating particular sub-tasks are usually available. Thus, for example, the analysis of a specific situation in need of planning can—depending on the task at hand and the amount of available time—be carried out by way of personal inspection, analysis of secondary data, one's own empirical investigations, or a combination of all three.

In the English-language technical literature, the rational model is also termed the "synoptic" (see for example Hudson 1979, 388) or the "rational-comprehensive" model of planning (see for example Sandercock 1998, 169).[4]

Relying on Herbert Simon (1947, 67) and Talcott Parsons (1949, 58), Meyerson and Banfield define the term "rational" (in the sense of "rational decision") as follows:

"By a *rational* decision, we mean one made in the following manner:

1. The decision-maker considers all of the alternatives (courses of action) open to him; i.e., he considers what courses of action are possible within the conditions of the situation and in light of the ends he seeks to attain;
2. he identifies and evaluates all of the consequences which would follow from the adoption of each alternative; i.e., he predicts how the total situation would be changed by the course of action he might adopt; and
3. he selects that alternative the probable consequences of which would be preferable in terms of his most valued ends." (Meyerson and Banfield 1955, 314)

According to Lindblom, "The hallmarks of these procedures ... are clarity of objective, explicitness of evaluation, a high degree of comprehensiveness of overview, and, wherever possible, quantification of values for mathematical analysis." (Lindblom 1959/1995, 36f)

Between 1945 and 1970, the rational model gained entry to many professional fields. Not only architectural and urban planners utilized the rational model, but it also found application in private ventures, in politics and in (non-urban planning related)

---

4   The term "rational-comprehensive" was most likely introduced by Charles Lindblom: "[this] approach ... might be called the rational-comprehensive method." (Lindblom 1959/1995, 37).

public administration. Special approaches such as operations-research, cybernetics and the so-called systems analysis helped to refine the model's methodological basis and tools. The rational model was so dominant in those years that Western methods of planning were all but equated with it: "… much of the 1950s and 1960s, Western planning thought became almost coterminous with the Rational … model …" (Weaver et al. 1985, 157 f). At the same time, the expectations of what this kind of planning could accomplish were considerable: The "soaring 60s" (Catton 1980, xii) followed the "golden optimism" (Catton 1980, xii) of the 1950s—with the aid of careful planning, it seemed at the time, no goal was beyond reach: "there are no problems, only solutions." (Catton 1980, xiii)

At that time, those planners who made use of the rational model assumed it was obvious that science and technology could be enlisted to make the world "function better." According to this model, then, the planner is an "expert" who relies on the "objectivity" of "professional expertise" to do what is in "the public's" best interest. He should be acquainted with the needs of the people for whom he plans, or, failing that, he should at the very least have the capacity to discover what those needs are. Furthermore, he has the additional task of working in the service of politicians: "Speaking truth to power", as Wildavsky put it in 1979. Planning was in vogue at the time. We all had faith that it was in principle possible to discover what was in the public's best interest. Terms like "public interest" and "the public" were rarely, if ever critically examined by planners, and for the vast majority of them "the public" implied an undifferentiated, homogenous group in which, for example, social, ethnic, or gender differences were seen as unimportant. At the time, most adherents of this model presumed that the agents of the particular planning boards not only have sufficient autonomy and authority to develop their designs through "rational" analysis, but also the power subsequently to bring these plans to fruition.

Disillusion with the rational model set in during the late 1960s and the early 1970s: planners realized the limits of what more technical and less socio-politically oriented models could accomplish. Correspondingly, dissenting voices that critically examined the model and its underlying assumptions, such as it was practiced at the time, began to mount. For example, the supposition that each individual step of the model (analysis of the situation, setting of goals, formulation of possible courses of actions to achieve those goals, and a comparison of the consequences of each course of action) is separate and independent of every other one was refuted. More important, though, was the discussion that erupted around disputed terms such as "objective", "rational", "optimal", or "expertise." It was ascertained that there is no such thing as "objective" knowledge, "rational" decisions, "optimal" solutions, and that all "expertise" is based on values and norms and thus operates on a more or less shaky ground. (See Lindblom 1959, Simon 1968, March 1978 and 1982, Alexander 1984, Popper 1987 or Mandelbaum, Mazza and Burchell 1996.) The model was therefore reproached for being: too positivistic,—meaning that it places too much faith in science and technology— ahistorical, and most of all, apolitical. The rational model basically ignored the fact that planning is influenced by norms and values— which is to say, politics—a great deal. "[T]he Rational … model … attempted to apply logical positivism to society. It defined rationality exclusively in terms of positive knowledge and instrumental calculation. Such knowledge was claimed to be objective and universal." (Weaver et al. 1985, 157 f) Its anti-democratic manner of planning "from the top down" was also criticized. Furthermore, many disagreed with this model because, to their mind, it encouraged acceptance of the *status quo*, support of the current political establishment, and maintenance of the norms and values held by the upper and middle classes.

Countless theorists have attempted to mitigate the rational model's shortcomings. The list of names extends from Herbert Simon (1976) with his concept of "bounded rationality" and "satisficing (rather than optimal) solutions", and Charles Lindblom (1959) with his strategy of "successive limited comparisons" or "muddling through", all the way to Amitai Etzioni (1967) and his attempt at "mixed scanning." In his combination of the two prior strategies, Etzioni accepted the process described as "muddling through" by Lindblom as a realistic description of how people actually plan in their day-to-day lives. At the same time, however, he proposes that Lindblom's pragmatic procedure for be adopted for more long-term planning as well.

An Explanatory Remark

At this point, an explanatory remark is in order. On closer inspection, the previous section manifests a contradiction: On the one hand, we criticized the rational model for the extremely technical, apolitical and ahistorical plans it elicited. Today there is no disputing that "mainstream" planning at that time was not only based on the rational model, but really was also extremely technical, apolitical and ahistorical in its execution. In this sense, then, critique of the rational model is certainly justified.

On the other hand, it is clear that the definition of the rational model—1. Analysis of the situation, 2. setting of goals, 3. formulation of possible courses of action to achieve those goals, and 4. a comparison of the consequences of each course of action (see Meyerson and Banfield 1955, or Muller 1992)—need not dictate these kinds of plans in actual practice. Plans based on the rational model need not always be overly technical, apolitical, or ahistorical. This is to say that the rational model was—indeed, at times still is (see Sandercock 1998)—accused of sins for which it does not bear the sole responsibility.[5]

It must be understood that the rational model possesses an unmistakable strength, namely, the fact that even its critics cannot avoid having to work through the same, or at least similar tasks. One way or another, they too must analyze the situation for which they wish to plan, set specific goals, formulate possible courses of action by which to achieve those goals, compare the consequences of each course of action, and finally settle on a course of action.

This difficulty lies in the fact that too many of the authors who currently publish on "Planning Theory" are content to only concern themselves edge-wise with the theoretical and practical implications of the task—i.e., analyze the situation, etc. As trivial as these implications may seem, the theoretical, methodological, and normative problems contained therein are anything but. Seni, for example, described the teetotalism of these planning theorists as follows: "However, the term 'planning

---

5   As Faludi (1996, 71) has shown, the imputation of fault has gone so far that the meaning of Meyerson and Banfield's (1955) text was obviously distorted in Lindblom (1959).

theory' belongs to a body of literature in public administration and environmental planning, and it would be wrong to assume that this broad and rambling literature ... provides the full context of what planning is in fact." (Seni 1996, 148) This is one reason among many why these days we so seldom hear any concrete suggestions for how to preserve those elements of the rational model that function appropriately while simultaneously overcoming those that have rightfully received criticism (see for example, Faludi 1987 or Schönwandt 1999).

To return to our seven models of planning, we will lay the concerns raised in this explanatory remark aside for the time being, anticipating that we shall pick the discussion up again in Chapter Two of this book (see also Schönwandt 1999 or Seni 1996). We will spend the rest of this chapter representing each model of planning just as it is discussed in the technical literature of planning theory. In so doing, we will only trace the main lines along which the planning-theoretical discussion has developed, even though there obviously exist a great many different approaches to planning[6] over and above what we will discuss here, and "any list of planning forms and styles could be extended almost indefinitely." (Hudson 1979, 390)[7] Any approach that lies beyond these main lines of development will not be considered beyond this point.

Let us—having concluded this brief explanatory remark—therefore return to the task at hand, namely to give an overview of the most influential models of planning that have been developed during the second half of the 20th Century.

The Advocacy Model of Planning

One of the first alternatives to the rational model came in the' form of the advocacy model. This model was developed in mid-1960s America as a reaction to the excessively technocratic and politically "top-down" approach of planning associated with the rational model. This new model became well-known with the publication of two articles: "A Choice Theory of Planning" (Davidoff and Reiner 1962) and "Advocacy and Pluralism in Planning" (Davidoff 1965).

This model is founded on the realization that "the public" is not a monolithic, homogeneous group; rather, it is best understood as made up of different interest groups. It was also recognized that power and access to resources is anything but equitably distributed among the members of this pluralist society. After all, some are richer and others poorer, some are well-educated and others less so, etc. Since questions concerning which goals planning theory ought to set itself, as well as questions about whose interests the so-called "public interest" really represents, don't simply fall under the domain of scientific planning (but are rather political questions), Davidoff urges planners to enter the political arena.

As a consequence of this view, advocacy planning demands the creation of *several* plans, each of which takes the various interests of different groups into account. This

---

6   For the difference between, e.g., "procedural" and "substantive" theories, see Faludi 1973 or Seni 1996.

7   "[I]ndicative planning, bottom up planning, ethnographic planning methods ... basic needs strategies" (Hudson 1979, 395) and so forth.

measure stands in contrast to the creation of a *single* (master-)plan, and is meant to follow from extensive discussions about the political values and interests that underlie each of the (several) plans. This serves to foreground the question of how to distribute limited resources ("Who is given how much and why?" as well as "Who receives preferential treatment, and who is left out?")—a question that, at least in its stronger formulation, tended to be ignored by those who availed themselves of the rational model until the end of the 1960s.

This model was meant to function in a manner analogous to the legal system: a lawyer (advocate, in this case the planner) helps "the weak" defend their own interests against "the powerful". This approach thus grows out of the conviction that decisions in planning are mostly influenced by the political lobbies of opposing interest groups. For example, it was argued that investors and homeowners possess the necessary influence to sway decisions in planning, whereas the poor who must rent their homes from the investors and homeowners do not. In these kinds of negotiations, the planner acts as a voice for the voiceless. He goes into poorer neighborhoods to discover what the section of the population who lives there wants and needs, and he supports the inhabitants of these neighborhoods by providing them with the expertise they require to assert their interests. In addition, he carries his findings back to the planning boards and city council meetings.

This model signaled a substantial step forward insofar as the interests of the disadvantaged and the underrepresented were explicitly taken into consideration, and that the norms and values that underlie planning were afforded more serious attention. In addition, more heed was paid to the side- and after-effects of planning, especially insofar as they created additional hardships for those who are already disadvantaged. In comparison to the rational model, a marked change in how planners actually go about their day-to-day work accompanies the advocacy model.

In actual practice, however, experiences with the advocacy model were rather more sobering. It quickly became apparent that the disadvantaged were not primarily lacking in the technical expertise and dexterity the planners had to offer, but rather in the power necessary to realize their ideas. Unfortunately, even advocacy planners didn't have at their disposal the power required successfully to represent the disadvantaged.

There were also other, less fundamental difficulties: It became increasingly evident that the analogy to jurisprudence—likening the planner to an advocate, i.e., a lawyer—left much to be desired. In political, as opposed to legal, debates there is no such thing as a independent third party to play the role of an impartial judge who, on the basis of pre-established laws and precedents, passes a verdict to resolve whatever conflict may have arisen. Generally, political conflicts can only be solved through compromise and politically mediated settlements. Hence, the work of advocacy planners was naturally called into question when necessary compromises failed to materialize. In actual practice, advocacy planners were accused of obstructing planning projects rather than offering useful alternatives.

In addition, it became evident that the idiosyncratic inclinations and preferences of individual planners play a deciding role in their work. Peattie, for example, described projects in which planners attempted to help the effected in order to lend greater political weight to their concerns. Peattie later characterized the work of these planners as the "manipulation-model" (Peatty 1968): the planners determined the political agenda, defined the problem as well as the terms in which the solution would be couched, and so forth. It became evident that planners concerned themselves primarily with those aspects of an assignment that would yield profitable results, rather than those that were most important to those who would be effected by the plan. Since the planners came from local planning boards, it was not rare for them to be more interested in their own careers than in the wishes and needs of their clients—a situation that often had the effect of leading citizens to avoid entering into conflict with the planning boards.

Robert Goodman is one of the most incisive critics of the advocacy model of planning. In his piece, "After the Planners" (1972), he describes planners as agents of social control, as "soft cops" of the political system. In Goodman's view, the incorporation of the disadvantaged into the planning process does not lead to any increase in their power—they are rather subjects of control and thus actually suffer a loss of power.

With the advocacy model of planning, Davidoff—with full faith in enlightened, pluralist democracy—developed a model that, while appearing to stand in contrast to its predecessor, could in actual fact be used to perfect it. The main problem with the new model was as follows: the concept of advocacy planning did not provide any concrete mechanism to dissolve actual disputes that arise between different interest groups. Though the model broadened the scope of the planner's role, it did nothing concrete to change the ruling power-structure. A core issue was therefore identified: should, or can advocacy planning actually go to the lengths needed to correct the unequal distribution of power and resources by, for example, proposing extensive ownership- or land-reforms?

As a result of their experiences with advocacy planning, various planners reached different conclusions about how to overcome the shortcomings of the advocacy model. They developed several models to replace the advocacy model, each of which will be described in the sections that follow:

(a) At the end of the 1960s and the beginning of the 1970s, a model of planning surfaced in reaction to the (neo)Marxist analysis of the connection between planning and capitalist society: the "(neo)Marxist model."

(b) Several planners, such as Norman Krumholz and Pierre Clavel, were convinced of the advocacy model's potential and attempted to improve on it. They suggested that planners ally themselves with like-minded politicians. Thus was born the model of equity planning (Krumholz 1994, Clavel 1994).

(c) Others began to rethink the role of planners anew. They shifted the focus of their work and concentrated on the process of the emergence of plans. This lead to the "Model of social learning and communicative action."

(d) A fourth group went so far as to drop the label of "advocacy" altogether. They turned their back on planning administrations and completely switched over to the side of the disadvantaged and underrepresented. They practiced the so-called "radical model of planning."

(e) At the same time, a fifth group urged planners to plan less altogether and give free reign to the open market instead. This approach is known as the "liberalistic model of planning."

The (neo)Marxist Model of Planning

Towards the end of the 1960s and the beginning of the 1970s, a planning model emerged in "capitalist" countries in reaction to the (neo)Marxist analysis of the structural relationships between planning and capitalist society. Two examples of this line of thought are Henri Lefebvre's *Le Droit à la ville* (1968, 1972) and Manuel Castells' *The Urban Question* (1977). According to this view, planning was and is first and foremost a political activity within a capitalist state. Correspondingly, the planner is no longer seen as an expert or specialist, but rather as a handmaiden of capital with rather naïve ideas concerning actual power relationships—power relationships in which he himself is deeply and inescapably imbricated.

This (neo)Marxist line of thought, with its sustained critique of so-called bourgeois planning as a function of the capitalist state, stood in the limelight at many university planning departments for several years. In a highly regarded case study on the development of a French city, Manuel Castells delineated three functions of planning (Castells 1978); according to this study, planning was an instrument

  (a) of rationalization and legitimatization
  (b) of negotiation and mediation between the differing demands of the various
      groups of capital, and
  (c) a regulator or valve for the pressure and protest of the governed classes.

No matter what the object of analysis of the (neo)Marxist theorists was (whether production, or consumption, or the role of the state in relation to the accumulation of capital or the distribution of goods etc.), their conclusions concerning the function of planning always remained the same: planning exists primarily in the service of capital, and the hope of effecting a change in this situation is illusory as long as the system remains as it is.

On the one hand, the emergence of this line of thought was a clear challenge to the traditional schools of thought in the field of planning. On the other hand, it only served to enhance the latent gulf between theory and praxis. Over the course of time, it became increasingly clear that the (neo)Marxist mode of thought appeared to have a paralyzing effect on political debate.

The value of the (neo)Marxist model is therefore to be found on the level of theoretical critique rather than on that of concrete planning. In particular, this theoretical approach once again called the concept of "public interest" into question, and made it clear that class interests are the true driving forces. In any case: a weakness of this approach was that it offered no new definition of the task of the planner, and no directives concerning what planners ought actually to do—other than enter into the class struggle. General answers like "the planner can become a revealer of contradictions and thus an agent of social innovation" (see Castells 1978, 88) were too weak to sufficiently inspire a generation of planners.

The Model of Equity Planning

The advocacy planners of the 1960s operated primarily outside councils and city administrations. Decisions, however, were ultimately still reached within the administration. For this reason, others attempted to pursue a path within the administration: their approach was to work with other like-minded and, in their view, progressive politicians in order to produce greater justice for the disadvantaged. Their thesis was: the city administration is an arena in which a political agenda is debated, and planners are particularly capable of achieving something if they fight on behalf of the interests of the disadvantaged *within this arena*. In Germany, this procedure was only rarely explicitly described, yet all the more frequently practiced.[8] The most prominent representatives of this model in the USA are Norman Krumholz, chief planner of the city of Cleveland, and Robert Mier, planner of Chicago.

The proponents of this model consciously tried to redistribute power, resources and the possibilities of participation away from the powerful and towards the disadvantaged and underrepresented. A comprehensive report about this type of planning is offered by Krumholz in his *Making Equity Planning Work* (Krumholz and Forester 1990), in which he describes not only the opportunities, but also the restrictions of this approach. At the bottom, this approach, like advocacy planning, assumes that planning doesn't operate in *opposition* to official policy, but rather in concert with it. The practitioners of equity planning do, however, consciously choose the politicians with whom they want to work. At its core, this model assumes—as does the rational model—that the planner is an expert and stands at the center as primary agent. However, while the advocacy planner primarily works in residential areas so as to discover what the local population desires and needs, the equity planner concentrates on the political arena. The advocates of this model stress the importance of dialogue, inside as well as outside of the administration: of course these planners also speak with those who are affected, but primarily they give interviews and write speeches for mayors, members of city planning boards and council members. They collect information, and analyze as well as formulate the problem at hand. They are, in other words, communicators and indefatigable propagandists. In this capacity, they have the power to guide attention towards specific issues and shape the discussion according to their own liking. On the other hand, those who utilize this model are nonetheless engaged in "top-down" politics—and thus in "speaking truth to power."

For the planner, several risks are associated with this approach: if, for example, the ruling majority of a city should change, then the planner has to countenance the possibility of being transferred into an insignificant department, thereby being "put on ice" and made inefficacious. In the USA, such a change may even result in the loss of one's job. This is to say that planners must be mobile, since they can only work effectively so long as they enjoy the support of the currently governing body of the city.

---

8   This assertion is founded on the author's own extensive experience in the practice of planning.

The Model of Social Learning and Communicative Action

With his idea of advocacy planning, Paul Davidoff attempted to effect an opening of the political process to those affected by planning, and thereby proclaimed a state of competition between various plans. To this end, the first advocacy planners put their (technical) know-how and services at the disposal of the disadvantaged. Over the course of time, however, some advocacy planners learned a different lesson—one described by John Friedmann in his 1973 book *Retracking America*—eventually leading to the model of social learning and communicative action.

These planners not only noted that the inhabitants of the planned neighborhoods clearly possessed the requisite technical abilities, but also became keenly aware of the gulf that existed between the so-called experts and their clients—a gulf which was only exacerbated by incomprehensible technical jargon. Faced with this conflict between the expert knowledge of the planners and the personal, lived knowledge of the inhabitants, advocacy planners concluded that neither of these two parties had *all* the answers. By way of a solution, they suggested bringing the two parties together in the context of a learning process—one of mutual complementarity in which the problems at hand would be worked through together. Friedmann termed this procedure the "transactive style of planning". Characteristic for this style of planning are dialogue, reflection of values, and mutual acceptance. In the course of this process, styles of thought and moral evaluation begin to converge; furthermore, mutual empathy creates a possibility for conflicts to be resolved in concert.

Resting on the same basic observation, namely, that planning is an essentially interactive, which is to say, communicative activity, a further line of thought developed in the 1970s and 1980s which defined planning as communicative practice. Inspired by the work of John Forester (see Forester 1989), and relying on Habermas' concept of communicative action, this group favored a model of "communicative rationality" in place of the rational model in planning. John Forester termed his theory "critical planning". For him, planning is primarily concerned with asking questions and listening critically in order to learn collectively in the context of a dialogue, thereby "sharpening one's own attentiveness". A deciding factor is the self-reflectivity of the planner who must scrutinize how he deploys his own power. He is interested in the "stories" which are told about the planning situation—only thus do the relevant power relationships, and not just the factual relationships, become visible. In this manner it becomes apparent which alternatives are even possible, and how these in turn may be realized in circumstances at hand.

The Habermasean approach employed here derives from the so-called critical theory of the Frankfurt School (see e.g. Habermas 1981, 1983). This theory assumes that science and scientific methodology don't produce simple, unassailable "Truth".[9] Rather, science must be seen as a tool which can be implemented in a process of manipulation, and which is therefore marked by the distribution of power in society. It can veil truth as much as it may reveal it: while a reality or truth may exist "out there", it is, for us, concealed behind socially constructed networks of consensus (in the form of e.g. assumptions, theories). These networks of consensus represent the

---

9   See especially Feyerabend (1975/1979), Kuhn (1962/1981), Toulmin (1972/1978).

power relationships in society, dominating our experience and thereby blinding us to other, alternative or "deeper" "realities". This basic thought has as its consequence that critical theorists espouse a conception of science opposed to the principle of so-called value-neutrality in science. Accordingly, the separation between politics and science propounded by the thesis of science's value-neutrality can no longer be maintained (see Kambartel 1996, 270). Habermas consistently bases his work on an idiosyncratic understanding of "truth"; he thereby tries to answer not only questions about truth, but also questions as to the justification of interests and norms. According to Habermas, the substantiation of (scientific) assertions as well as the justification of norms may be established by way of a universal accord arrived at through non-coercive means—a pragmatic concept of "truth": what is true is what participants in an unconstrained communicative or ideal conversational context ("the ideal speech situation") will accept as true. The essential criteria for such an ideal conversational context are that every participant of the group be in possession of the same information and that all points of view be represented. Furthermore, the conditions must be such that the power of the argument, and not the power of a given individual, may be the decisive factor. This form of life and manner of knowledge creation is termed "communicative rationality" by Habermas. Validity and truth are thereby yielded through rational argumentation within a discourse: the strength of an argument is determined in accordance with its capacity to convince participants within a discourse—that is, according to whether it prompts them to accept the validity of the assertions. The only power active in an ideal speech situation, or within communicative rationality, is therefore the "power of the better argument".

For planners, the idea of communicative practice is attractive because this approach doesn't require planners to search out a value-neutral expert-role. It offers the possibility of orienting oneself towards normative, political foundations. It contains an encompassing social component and offers extensive freedom to determine the content of planning-oriented work oneself. The emphasis doesn't lie so much on what planners know, but rather on how they utilize and distribute their knowledge; less on their capacity to solve problems, and more on the question of how one may open and direct debates on particular subjects. This model thus concerns itself with speech, argument and the sharpening of ones attentiveness.

The novelty of this approach consisted primarily—as described above—in a new understanding of expert-knowledge, as well as of knowledge in general. The monopoly of the expertise of professionals was dissolved, and the value of lived, local knowledge was recognized. Furthermore, in place of the concept of fixed, static knowledge a dynamic concept of knowledge and learning was favored. It was, in other words, no longer a matter of carefully analyzing the inhabitants and defining their needs and desires, but instead intensively involving them in the process of planning and incorporating their knowledge and abilities. The theory of communicative action was an attempt at strengthening the democratic aspects of planning and knowledge creation—at removing communication barriers and establishing an open discourse.

Upon closer inspection, the model does, however, betray certain weaknesses— ones that inhere less in the model itself than in Habermas' theoretical framework. For one, Habermas' criterion of truth (see above) harbors the same threat of conventionalism that earlier thwarted the so-called coherence theory of truth (see

Groeben and Westmeyer 1975, 142ff). Truth cannot be determined via convention or agreement alone, because conventions are arbitrary and thus not a sufficient criterion of truth. In his discourse theory, Habermas also fails to distinguish between knowledge (e.g. the contents of concepts, assertions, theories etc.) and the social determinants of the construction of the contents of knowledge. As a result, the planners who employ this approach primarily concern themselves with the social determinants of knowledge, leading them to neglect work on conceptual content and constructs (see part two of this book). Furthermore, group-dynamic processes may develop a dynamic of their own; for example, the willingness of those involved in the planning process to enter into risks may increase under certain circumstances, or an overly hasty harmonization of judgments concerning a state of affairs may occur (see Janis 1972, Hayes 1996, 158ff or Brown 1997). Finally, the very concept of an "ideal speech situation" is criticized because such situations simply don't occcur in reality. Flyvbjerg has set forth the basic conditions framing this conversational context: truth, validity and consensus are guaranteed if the participants in a discourse respect five procedural demands of the discursive ethic:

(a) No party affected by what is being discussed should be excluded from the discourse (the requirement of generality);
(b) all participants should have equal possibility to present and criticize validity claims in the process of discourse (autonomy);
(c) participants must be willing and able to empathize with each other's validity claims (ideal role taking);
(d) existing power differences between participants must be neutralized such that these differences have no effect on the creation of consensus (power neutrality); and
(e) participants must openly explain their goals and intentions and in this connection desist from strategic action (transparence). (Flyvbjerg 1998, 188).

This list clearly demonstrates the limits of this approach. While Habermas certainly recognizes the problem of power and structural inequality within society, he ends up bracketing them by proclaiming a discourse free of power. There is, however, no space free of power, certainly not in planning. Habermas desires an ideal situation, but fails to provide directives for how we might actually achieve one.

Against this backdrop, recent work explicitly engages the topic of power. In so doing, it sees power not only as a negative, but also as a productive and constructive force. Power is not simply concentrated at centers. It is not something one can possess, but must rather be understood as a dense net of heterogenous relations. One central question is therefore not so much who possesses power and why, but how power is exercised—the focus thus lies more in the process than in the structure (see Flyvbjerg 1998, 185ff or Foucault 1982, 210ff).

The Radical Model of Planning

Some planners extracted a different moral from the difficulties of the advocacy model (see e.g. Heskin 1980). Advocacy planners generally found themselves trapped in an insoluble dilemma: as employees of planning administrations, they couldn't seriously fight against their own planning boards without having to fear negative repercussions for themselves. It was therefore difficult for them to commit themselves unconditionally in support of the interests of those ultimately affected by the plan in question. If, that is to say, they really wanted to achieve something on behalf of the disadvantaged and against the inequitable distribution of resources, power etc., they could no longer remain employees of the planning administration. Success in this struggle was, in other words, only possible from outside this administration, often even in active opposition to it.

With exactly this precept in mind, a new approach to planning emerged: the radical model of planning. The core points of this model are essentially as follows: just like the proponents of the model of social learning and communicative action, radical planners turn themselves against the domination of expert knowledge; they are open to the possibility of learning through action or experience, and recognize the value of knowledge thus achieved. Unlike the former, however, they act in clear opposition to governmental organizations or economic interests, or both. They, at least partially, reject the political systems of society and try instead to change economic and political structures with a view towards eliminating systemic inequities. Consequently, radical planners abjure traditional methods and parliamentary procedures, preferring to work outside the system instead. This model was applied in many areas: the spectrum ranges from homeless initiatives, to urban planning, all the way to ecological parties/ organizations and peace movements.

Radical planners thus assume a new role—one that requires a new understanding both of what it is that planners do, and of what it means to be a planner. Nothing short of a new professional identity is involved here. This identity includes, among other things, that, instead of working within the context of a professional community, the planner must relinquish his former status in order to lend his support to the disadvantaged. A radical planner cannot cling to his professional environment and simultaneously hope to support those who are most affected by planning. In order to help them, he cannot see those affected by planning as mere "clients", but must instead become a part of, or at least assume solidarity with, the group in question.

In concrete terms, the emphasis of this approach lies primarily on the kind of communal action best suited to yielding tangible results as quickly as possible. The political system is generally not directly confronted in the process. Instead, attention is (enthusiastically) focused on those areas that remain beyond the reach of the system.

Many of the initiatives undertaken according to this model doubtlessly provided decisive stimuli and yielded formidable results. The numerous successes of the ecological movement can certainly be counted among them. Several problems are nonetheless associated with this approach: one of the primary difficulties is that the advances effected through radical planning often quite quickly encounter legal and financial barriers, especially if anything beyond marginal innovations are aimed at. A change in the tax structure of private energy consumption or passenger-car traffic, for example, cannot generally be realized from a position outside the political system. If such barriers are to be broken, suitable initiatives within the political system are necessary—new laws must, for example, be enacted, or existing ones changed. In the end, this means that within societies such as we know them, planning initiatives based on this model cannot represent anything but a transitional stage if they are to achieve more than small, piecemeal victories. If radical planners do achieve some form of success in their struggle against the status quo, they are generally not able to avoid subsequently becoming part of the very system they had hoped to change.

To this, a further problem must be added: with their switch to "the other side", radical planners were able to create clear and easy distinctions between the opposing fronts. "We" are here, "the others" there. They thereby not only oversimplified their own relation to the state, but also idealized the very concept of the "community" they intend to serve. The community of the "disadvantaged" or the "underrepresented" is not a homogenous group. People often belong to many, even very different, groups at the same time; these allegiances , furthermore, tend to shift from time to time: there is no shortage of groups that attempt to use planning in order to block others from accessing their privileges and resources: natives versus foreigners, the employed versus the unemployed, various religious groups against each other etc. In the end, this means that the neat and enduring separation between "us" and "them" postulated by radical planners cannot generally be maintained in pluralistic societies.

Beyond this, it should be noted that the radical model's ability to function depends heavily upon group-size and the number of agents involved: as the ambitions and activities of such movements grow, they (just like all traditional planning administrations) tend to require clear delegation of responsibility, formal organization, as well as the attendant hierarchy—which, in turn, tends to manifest the very weaknesses that are criticized in the existing system.

The Liberalistic Model of Planning

In the context of the liberalistic model of planning the word "liberalistic" stands for "laissez-faire". At the bottom, the proponents of this model assume that planning ought to intervene only if the mechanisms of the "free market" have failed and planning can be seen to be clearly preferable to letting things "go their own way" (see e.g. Sorensen and Day 1981, Sorensen 1983). Instead of relying on planning to protect people, nature etc., the proponents of this model place their trust in individual (property) rights, the interests of individuals to maximize their own well-being, and the power of contracts that people enter into amongst themselves.

In this model, planning serves the purpose of supporting and expanding the freedom of action and the possibilities of self-realization available within the framework of the free market; of protecting the rights of individuals and regulating undesirable consequences produced through the behavior of the individual; and of providing compensation for infringements against individual rights. Behind all this stands the maxim of several economic and economico-political theories: as little planning as possible, and only as much planning as necessary. The use of resources in the service of planning is thus seen as a necessary evil, to be avoided whenever possible.

The liberalistic model possesses a variety of strengths. Its proponents often point to prevailing and unreasonable expectations as to what may be achieved through planning; they warn, not without justification, against an exaggerated "mania of planning and regulation" and proclaim "deregulation" in its stead. The weaknesses of this approach are, however, considerable—the concept of the "free market" implies scores of difficulties. For one, the freedom of this market is relative because it functions according to numerous explicit and implicit rules. Furthermore, this market is only really free to those who satisfy certain initial requirements: financial means, requisite knowledge, time etc. To all others, entry is barred; which is to say that the goal of protecting the rights of the individual (see above) is only conditionally realized. Thus, while the liberalistic model of planning does incorporate a concept of "freedom", it simultaneously ignores concepts like "equality".

A further weakness of the liberalistic model is its heavily circumscribed concept of planning, in which planning is really confined to "public planning" (see Sorensen 1983). It thus overlooks the fact that even private enterprises must engage in planning to a considerable degree in order to survive in the "free market"—something that indeed couldn't be otherwise given that any form of responsible management of resources, be they public or private, presupposes planning.

Summary

In this chapter, we discussed summary overviews of seven different models of planning. They are:

(a) The rational model of planning, which was criticized for being too positivistic, apolitical and ahistorical.
(b) The advocacy model, which moves "the disadvantaged", "the underrepresented", and thus the question of distribution ("who receives what and how much, and why?" and "who is privileged and who is disadvantaged?") into the center of our thinking, and therefore requires planners to draw up several different plans rather than a *single* (master-)plan.
(c) The (neo)Marxist model of planning, with its attendant ideology.
(d) The model of equity planning, which proposes planners ally themselves with like-minded politicians.
(e) The model of social learning and communicative action, in which the planner is no longer an "expert". Rather, the practical knowledge or lived knowledge of each inhabitant is acknowledged and it is understood that both parties (i.e., planners and inhabitants) can learn from one another. Consequently, an accordingly important role is played by dialogue, reflection on each other's values, and mutual acceptance.
(f) In addition, there was the radical model of planning, in which the planner turns his back on administrators, as well as
(g) the liberalistic model of planning, in which planning ought to be minimized altogether and as much as possible left to the mechanisms of the "free market".

Most of these models have their own collection of methods, data requirements, professional abilities and styles, as well as their own institutional milieus. Furthermore, in actual practice they are often used in conjunction with one another.

Chapter 2

Towards a "Third Generation" of Planning Theory

Introduction

Planning theorists have been in search of a new, synthesizing approach to describe their field since the late 1960s and early 1970s. As we saw in the previous chapter, many authors of that era argued that it was time to abandon the "rational" model,[1] which constitutes the "first generation" of planning theory. But what was to take its place? Since then, have the products of their effort coalesced into an approach that maximally takes into account and efficiently integrates every aspect that plays a role in planning? The answer is "no". The state of the discussion in the middle of the 1980s begged the obvious question—"After Rationality, What?"—that Ernest Alexander posed in the *Journal of the American Planning Association*'s 1984 issue. Twelve years later, he characterized the situation as follows: "Planning theoreticians are in a state of turmoil. Nothing is accepted; everything is questioned." (Alexander 1996, 45)

This section of the book will, after having described some earlier theories, introduce, in outline form, a planning theory that we hope does justice to as many of the complexities inherent in the planning process as possible. This theory encompasses the substantive (meaning spatial, social, political, ecological, and economic) elements of a given planning task. It takes into account the restrictions on our perceptual and cognitive abilities, as well as the degree to which planners are capable of controlling the course of events as they unfold. In addition, it connects with some relevant aspects of the theoretical background, especially the semiotic, epistemological, and ethical dimensions of planning.

Three Generations of Planning

The "First Generation" of Planning

The main task in planning theory and methodology is to analyze, compare, develop, and apply scientific theories and methods that can support the planning process in practice. The field's development from the end of the 1940s up to the 1970s can roughly be described as follows:[2] Immediately on the heels of the "golden optimism" (Catton 1980, xii) of the 1950s followed the "soaring '60s" (Catton 1980, xiii), with

---

1 In this chapter, "Towards a 'Third Generation' of Planning Theory", the terms "planning theory" and "model of planning" will be used synonymously.

2 An account of the field's history can be found in Friedmann 1996.

their overly confident estimation of what could be achieved through planning: "there are no problems, only solutions" (Catton 1980, xiii). Disillusion only set in during the 1970s: the limits of what could be accomplished using the more technical, less socio-political methods that were popular in planning at that time became apparent.

It was with this in mind that Rittel (1972), among others, described the gravest weaknesses infecting the then dominant understanding of how one ought to carry out the planning process. In doing so, he posited a distinction between two generations—the "first" and the "second" generation—that have dominated the various approaches to planning. According to the "first generation", that is to say, the generation which was practiced and taught until the beginning of the 1970s, the planning process is partitioned into the following eight phases:

- understand the problem
- gather information
- analyze the information
- devise solutions
- assess the solutions
- implement a solution
- test the solution
- modify the solution, if necessary.

Various authors have called the steps by different names—sometimes subdividing them more coarsely, sometimes more finely—but the basic principles always remain the same.

This way of understanding how one ought to plan implicitly contains the following assumptions (see Rittel 1972, 390):

- The formulation of a problem and its solution constitute two discrete phases, and one is independent of the other.
- The approach should be "rational" as well as "objective".
- Not only one specialized discipline should be involved in the process, as planning ought to be an interdisciplinary effort.
- The solution should be "optimized", meaning that all the relevant aspects that are to be maximized should be unified into one single measure.

This point of view forms the basis for the "rational" model in planning. The image of a rational agent—an agent who makes decisions based solely on the basis of rational deliberation and sound criteria—reigns supreme in this model. The imagined agent is informed of all the alternative solutions available to him, and chooses the solution that promises to maximize his utility on the basis of purely rational criteria. The assumptions that must be met in order to deploy this model are largely as follows (see, for example, Lindblom 1959, Mayntz 1976, March 1982, Fredrickson and Mitchell 1984):

(a) Comprehensive information about

- the features of the planning assignment

- the alternative solutions
- the consequences of each alternative solution
- the assessment of each alternative (through numerical values), in terms of the relevant features that are important for the planning assignment.

(b) Unequivocal goals and desires, which exhibit the following qualities: they are

- stable over an extended period of time
- independent of the other alternatives to be assessed
- non-conflicting, or at least comparable,
- and it is possible to arrange them in order of their importance, irrespective of the current planning situation.

Even a cursory glance at the above list clearly indicates how unrealistic (and therefore inapplicable to most actual planning assignments) these assumptions are. Hence, the rational model of planning has rightly elicited criticism on more than just one occasion (see the first chapter of this book or, e.g., Lindblom 1959, Simon 1968, Rittel 1972, March 1978 and 1982, Alexander 1984, Popper 1987, or Mandelbaum, Mazza and Burchell 1996).[3]

The "Second Generation" of Planning

Rittel therefore contrasted this "first generation" of planning with what he termed the "second generation" of planning. According to Rittel, the "first generation" characterized planning as the working out of "tame" problems. The "second generation" was markedly different: it proceeds from the assumption that, in planning, one usually has to deal with "wicked" problems (see Rittel 1972, Rittel and Webber 1973). Examples of tame problems include a game of chess, or solving a mathematical equation. In problems such as these, the task at hand, the admissible avenues by which one can reach a solution, as well as the objective one hopes to obtain are all defined clearly and unequivocally. So, for example, there are clearly defined rules we must all follow in order to play chess or solve mathematical equations properly. Likewise, it is clear when a game of chess has been concluded or a mathematical equation solved (at least for those of us familiar with the aforementioned rules).

But the so-called "wicked" problems behave differently. (Other authors have described them as "ill-defined problems" or "ill-structured problems"; see, for example, Simon 1973). They are primarily characterized by the following properties:

---

3 The rational model represents what was the "mainstream" of planning theory at that time. Of course, there were also authors who made use of a different, more socio-politically oriented approach to planning (see especially Meyerson and Banfield 1955). This is reason enough to assent with Faludi's thesis: "The idea of objective rationality is wrongly imputed to advocates of rational planning by their opponents ... So, by claiming that rationality purports to transcend conflict, its critics have created a straw man." (Faludi 1996, 71) All the same, planning theorists have found themselves intensely engaged by the rational model for the past three decades (this is true under the assumption that we include here the search for an alternative to this model).

(a) Each wicked problem is essentially unique.

(b) There are no finalized definitions in wicked problems. Each description of a wicked problem is only provisional and can be understood as a symptom of yet another problem.

(c) The discrepancies that obtain in various representations of wicked problems can be explained in any number of ways; the choice of an explanation determines the method of solving the problem at hand.

(c) Wicked problems have neither an enumerable (or fully describable) set of potential solutions, nor is there a clearly defined set of measures that may be incorporated into the planning process.

(e) Solutions to wicked problems are neither "right" nor "wrong", but rather "better" or "worse".

(f) Planners cannot conduct experiments, as do scientists in the lab. Thus, they cannot afford to make mistakes.

(g) For this reason, every solution to a wicked problem must be a "one-shot-operation". This means that every attempt is significant and final, as there are no opportunities to learn by trial and error.

(h) Since all definitions and descriptions of wicked problems are necessarily provisional (see above), there are no rules according to which planners can know to have completely solved them.

(i) There are no immediate or final means by which to verify the solution to a wicked problem.

Undoubtedly, the Rittelian "second generation" initiated a whole series of changes in planning theory. Most important among the effects of these changes is the fact that planning was thereby made to stand on a more sophisticated philosophical foundation. This manifested itself in the acceptance that knowledge is never wholly reliable and that it always depends crucially on certain metaphysical assumptions or paradigms (see, for example, Kuhn 1962/1981, Toulmin 1972/1978 or Feyerabend 1975/1979). This resulted in planners realizing that concepts such as the "objective" description of a problem, or the "optimal" solution to a problem, could no longer serve as the basis of discussions in planning (nor, indeed, should they so serve), because neither really exists in the first place. Moreover, the significance of values (axiology and ethics) became apparent to planners.

The shortcoming of the Rittelian program is that it ignores many of the aspects that come up in actual planning tasks. Consequently, this "second generation" does not offer a systematic survey of the field. Indeed, a systematic approach would contradict the basic philosophy that underlies this "second generation"; after all, a proposition such as "each wicked problem is essentially unique" (see above) requires the assumption that nothing is transferable in planning. It follows from this that—to put it bluntly—nothing can be systematized in planning. As Alexander (1992, 9) has correctly pointed out, Rittel's "second generation" is therefore based on the notion that it is impossible to formulate a systematization of planning theory (see also Mandelbaum 1979).

Responses to the Breakdown of the Rational Paradigm

Rittel was not the only one who voiced his opinions concerning planning theory in the early 1970s. At the same time and in the years that followed, there were many attempts to further the discussion about what could take the place of the rational model. Were one to provide a retrospective summary of the results that this work engendered, one might arrange them as follows (see Alexander 1984; 1996, 47 ff):

1. The *ritual response* is relatively widespread. It urges that we hold fast to the rational model. A study conducted in the United States during the 1980s showed that over half of all university planning departments still taught the rational model.
2. The *avoidance response* ceases to search for an integrative model of planning. One instead either concentrates on describing how planners actually behave, or one slightly modifies the rational model.
3. The *abandonment response* most often follows from the assessment that the rational model—or rather, anything on or above the rational model's level of abstraction and generality—is either unnecessary or impossible to implement (see Rittel and Webber 1973, Mandelbaum 1979), and ought therefore to be avoided altogether.[4] One alternative of this abandonment response postulates (at a clearly lower level of abstraction) that the intuition or experience of the planner forms the true basis of action. Or, it is sometimes argued that the job of a planner is only of a pragmatic nature, and that it ought therefore to be of a very limited scope anyhow. According to yet another version, the "value-neutral" rational model is replaced by concrete ecological, technological, social and political ideologies (for example, neo-Marxism, ecological fundamentalism, Cargoism (see Catton 1980) etc.).
4. The *search response* constitutes the fourth reaction to the rational model. Its basic approaches can be divided into two main groups. On the one hand, the topics of the first group are somewhat *too* general for the present context: although they deal with planning as a whole and are therefore relevant to planning, they fail to differentiate between individual parts of a planning task. We may count among these the action theories (to describe the procedures of planning), ethics (norms and values as the basis for every decision in planning), or the so-called communicative theories of planning (for the development of these norms and values),[5] etc. The work of the second group, on the other hand, concentrated (excessively) on individual aspects—example: "Planning is communication" (Selle 1997, 40)—or certain combinations of individual aspects of a planning task, that are then either discussed as opposing poles (example: positivistic versus normative theories; see Yiftachel 1989, 24ff) or dealt with dialectically (example: thesis: an object-centered point of view, antithesis: a subject-

---

4   More than a few planning theorists altered the main focus of their work and began to stress historical aspects of planning.

5   See, for example, Forester 1989, and Habermas 1981, 1983.

centered point of view, synthesis: transactive planning; see Banai 1988, 15ff). Or two poles are crossed with each other to form a four-part network: substantive versus procedural plus explanatory versus prescriptive theories (see Yiftachel 1989, 26f). Or five groups are formed right from the start, as we see in Hudsons SITAR model (synoptic, incremental, transitive, advocative, radical; see Hudson 1979, 388ff)—and so forth.

All the above-mentioned themes are relevant and helpful; they add precision to significant individual aspects or specific portions of a planning task. What is currently missing, however, are theories situated somewhere in between the (excessively) general approaches (of the first group, see above) and the (excessively) individualized approaches (the second group, see above)—theories, in other words, that conclusively integrate as many of the aspects that come up in planning as possible and systematically connect them to one another. Theories of this type will be characterized as planning theories of the "third generation". These theories ought to serve as constructs that guide our actions in planning (see Part II of this book), and in the absence of these approaches relevant aspects of a planning situation are easily overlooked.

Outline of a "Third Generation" in Planning Theory

In this section of the book an attempt will be made to move planning theory one step further into the future. The model described in what follows was chiefly anticipated by Heidemann (1992, 14 and 95; 1995), though only the first of these sources (1992) has been published.

In order to structure the complex subject of "planning", we will initially fall back on systems theory, with the caveat that we must ask forgiveness if, for the sake of our present purposes, we have over-simplified matters somewhat. (On the subject of systems theory, see, for example, the introductory texts by Jantsch 1992, Siegwart 1996, or Luhmann 1996). Here there are primarily two different approaches. According to the conventional definition, systems are chiefly understood as networks of components and relations that together integrate parts into a whole; systems are "a set... of ... [components], such that an interrelation obtains between them. Examples include an atom as a system of elementary particles, a living cell as a system ... of innumerable organic connections or enzymatic reactions, a human society as a system of many individual people that are in diverse relationships with one another." (Huber 1976, 6) This is the definition of the so-called component-relation model of systems. However, this model has not escaped criticism. "The inadequacy of this conception of systems lies largely in the fact that it isolates the system unto itself by focusing exclusively on its "inner workings"—that is, on the ... [components] and their relation to each other and to the whole—without any regard for the environment." (Bäcker 1996, 68)

The System-Environment Paradigm of Systems Theory

The so-called system-environment paradigm circumvents this shortcoming. According to this paradigm, systems are made up of a system-core, which is embedded within a larger environment. The collection of all that makes up the system's core—which are components and relations—is called the composition of the system's core (see, for example, Bunge 1979 or Mahner and Bunge 1997). However, this composition is supplemented by all those numerous things that are not part of the system's core, meaning the system's environment. This approach is based on the assumption that systems theory must of necessity be a theory of the relationships that obtain between the core of a system and its environment (see, for example, Luhmann 1996, 35). The reason for this is that cores of systems are not only sometimes, but rather structurally, and hence always, tied to their respective environments, and therefore can't exist without this environment. A system is therefore always a "core-of-a-system-within-an-environment" (see Bäcker 1996, 67).

Of course, when modeling a specific system it is usually not necessary to take the entire universe into account as environment. Rather, it suffices to include those portions of the universe in the environment that either influence the core of the system or that are themselves influenced by the core of the system. The term "environment" is therefore not to be understood holistically, such as it was for Plato, the Stoics, or Hegel, because we cannot make a distinction between a directly relevant and a generic environment. Only the collection of those things outside of the system's core that are somehow tied to the core are to be understood as the system's environment. That is, we will here use the term "environment" exclusively in the sense of a "direct or immediate" environment (see Mahner and Bunge 1997, 25).

There is no Planning "per se"

The following thesis underlies all of this section: there is no planning "per se". Planning is always carried out by people who possess specific biological and psychological idiosyncrasies, who nearly always interact with organizations or cooperative projects, who live and work in their own social and cultural milieu, and who have certain aptitudes, skills, and shortcomings, which is to say restrictions. For example, these restrictions include the fact that our perception of the world is in principle selective, meaning incomplete, and that it is necessarily influenced by cognition and is therefore always theory-laden. To this we must add the limitations that underlie our cognitive abilities as well as the fact that we do not have the power to change all that we may wish were otherwise than it currently is. A model for planning should encompass all these qualifications and limitations. For those in search of such a model, the approach pioneered by Jacob von Uexküll offers itself up as a suitable basic framework. This approach includes the above-mentioned systems-theoretical components and explicitly addresses not only the limitations placed on our perceptual abilities, but also the restrictions of our cognitive capacities and the limits of our ability to act.

The Functional Circle of Jacob von Uexküll

Today, the systems theory that follows the system-environment paradigm is occasionally called the "newer" systems theory (see Bäcker 1996, 67). But this overlooks its longstanding tradition: the system-environment paradigm is already contained within the so-called "Functional Circle" of the biologist Jacob von Uexküll (1928/1973).[6]

According to Uexküll, every organism—and therefore also us humans—is a subject that, due to its "design", is only receptive to some of all the stimuli produced by the external world, to which it responds in particular ways. These responses consist of specific effects on the external world, and these effects in turn influence the stimuli. A closed circuit is thereby formed, called the "Functional Circle" (see Uexküll 1928/1973, 158).

This "Functional Circle" can be articulated as follows (see Figure 2.1): The stimuli affecting the organism constitute that organism's "Sensory World", which it apprehends with the aid of a specialized "Sensory Apparatus". The "Sensory World" therefore extends only so far as to include those features of the "Environment" that the "Sensory Apparatus", which is to say the sense organs, of an agent can detect. Stimuli that cannot be detected essentially do not exist for an agent and are therefore also unavailable for interpretation by him.

The "Cognitive World" of an organism, which is its inner world, is the world in which it, using its "Cognitive Apparatus", grasps the possibilities, as well as the limits, of its capacity to influence or control a course of events as they unfold.[7]

The effect an organism, using its so-called "Effectual Apparatus", has on the world gives rise to that organism's "Effectual World". The "Effectual World" of an organism extends far enough to cover only those portions of the "Environment" that it as an "Agent" is operationally outfitted to affect: "Agents" can modify only those things that are within their reach.

An "Agent's" "Setting" consists of all the facts and processes within its "Environment" that are accessible to the "Agent" for observation and/or alteration. The "Sensory World", "Cognitive World", "Effectual World", and "Setting" are all

---

6　Many readers might be irritated by the decision to fall back on a model from the 1920s. Of course there have been more recent attempts to make useful planning models that bear some similarities to the Uexküllian approach (see Newell and Simon 1972, Faludi 1973 (for a criticism of Faludi see Thomas 1982) or Stachowiak 1992). However, none of these models contain the systems-theoretic components described above so explicitly as does Uexküll's approach. These models were often understood more as analogies to the way a computer functions, which is a way of thinking that has little to do with Uexküll's intention (see, for example, Bechtholsheim 1993).

7　Von Uexküll's functional circle applies to all living organisms, not only us humans. However, since the illustrations that follow are meant to apply only to humans, there is no need to justify the decision to deploy concepts such as "Cognitive World", etc. Of course, the unique standing of humans among all other living organisms has increasingly been called into question lately: "Pigeons behave according to the laws of formal logic, ... desert mice construct categories in the Kantian sense." (Halentz 1997, 60).

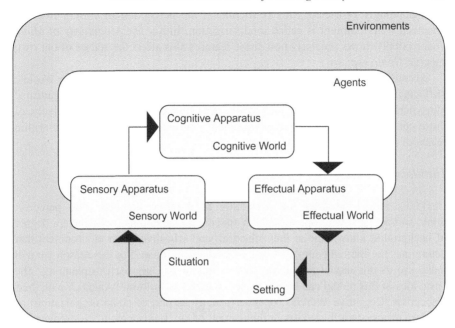

Figure 2.1   The functional circle

Source: Heidemann 1992, 14; (modified) after von Uexküll 1928.

related to one another recursively.[8] (For further details, see von Uexküll 1928/1973, 150 ff)

Uexküll propounded that all living organisms stand in a very specific relationship to their environment—which means that his theory mirrors the system-environment paradigm (see above). This elementary relationship consists of the fact that an organism lives in what he calls a specific medium, which, for the organism in question, is both self-evident and accessible only by way of its "Sensory Apparatus". The absorption and processing of the information available to the organism is also

---

8   Uexküll's functional circle is well illustrated by the extreme case of a tick (Ixodes rhicinus) (see Riedl 1980; Süskind 1985, 26f). The tick's Sensory Apparatus do not adequately equip it to engage with the totality of all the signals produced by its Environment. It only has two bits of information from its surrounding environment at its disposal: the odor of butyric acid and the perception of a narrowly defined range of temperatures. This definition of "mammal" in the tick's "worldview" (according to the presence of an odor indicating butyric acid accompanied by a definite range of temperatures) is impossible to beat, either in simplicity or in accuracy. "Error is all but out of the question." (Riedl 1980, 43) The odor of butyric acid causes the tick to let itself drop and, as soon as it perceives a temperature of 37 degrees centigrade, it begins to burrow underneath the mammal's skin. This is to say that, in the case of a tick, its actions in the world consist only in this, to "check" and make sure that both properties, butyric acid and 37 degrees centigrade, are present. If this turns out to be the case, then it activates the entirety of what is available within its "Effectual World", namely to let itself drop and burrow.

correspondingly restricted. This restriction in the perception of stimuli from an organism's environment is called modularization. In his *The Modularity of Mind*, Fodor (1979) shows precisely how these features also affect the senses of our own species, *Homo sapiens*.

According to von Uexküll, all four worlds ("Sensory World", "Cognitive World", "Effectual World" and "Setting") are marked and determined by an organisms' idiosyncrasies. Moreover, its immediate "Environment" consists exclusively of those components of the public environment that are accessible, meaning within reach, of the organism.

The Planning Model

Uexküll's model provides the possibility to describe planners who perceive, think, and act (and who operate within specific organizations), who have a degree of background knowledge at their disposal, and who live in an environment that determines the limited courses of actions available to them. For this reason we will make use of this model as a basic framework for our approach to planning. The basic idea of this model can briefly be summarized as follows: "Agents", with their respective "Cognitive Worlds", operate (usually as part of some organization) as the core of a system that exists in the context of an "Environment". Moreover, the "Agents" that make up system-cores are constantly engaging in exchanges between themselves and various components of their "Environment".

Adapting some Expressions

The next step toward our planning model goes as follows: since many of the terms that we have been using, such as, e.g., "Sensory World", "Effectual World", etc., are unfamiliar to planners, we will replace them with different expressions, ones that are likely to be more intelligible in the context of planning (see Figure 2.2 as well as Heidemann 1992, 95).

A deliberate effort has been made to use concepts (for example, "Comprehension of the Situation", etc.; see below) whose meaning is not too narrowly delimited. This is mostly due to the following two reasons: for one thing, many of the concepts commonly used in planning (such as "means", "goal", etc.) have become objects of controversial discussion (see, for example, Faludi 1987, Alexander 1992, Fischer and Forester 1993, or Dym 1994). These expressions therefore pose a problem and are not suitable for a model such as ours without being accompanied by additional explanatory elucidation. Secondly, the use of a deliberately broad terminology allows for the model to be applied to a whole range of different planning tasks.

The modified expressions are now as follows (see Figure 2.2): We will substitute the term "Planning World" for what was previously designated as an "Agent" (see Figure 2.1). Hopefully, this choice in nomenclature will serve to clarify that we are not only dealing with the individual people who operate within the "Planning World" (and usually cooperate within some organization), but that we must also take their background knowledge (viz. the methods, concepts, theories, worldviews, and paradigms on which they draw) into account.

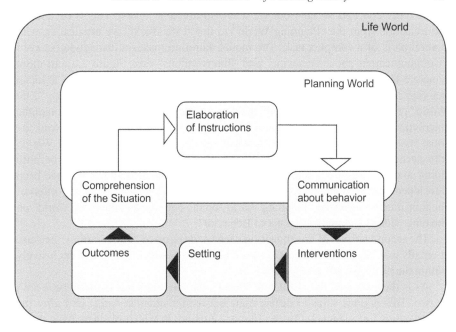

Figure 2.2   Basic scheme for planning

Source: Heidemann 1992, 95 (modified)

What was called an "Environment" in Figure 2.1 will be exchanged for the expression "Life World", By "Life World", we mean the environment of the system-environment paradigm, which sets the parameters, i.e. the background, for actions in the "Planning World". The "Life World" therefore lies outside the "Planning World" such that the latter is embedded within the former and such that both of them influence each other in a reciprocal manner.

The remaining terms are modified for Figure 2.2 as follows:

The "Sensory World" is translated into "Comprehension of the Situation". This refers to that portion of the "Life World" that the "Agents" of the "Planning World" can perceive and interpret. Furthermore, the "Comprehension of the Situation" builds a bridge between the "Life World" and the "Planning World". It is therefore one point at which the "Planning World" opens up to the "Life World".

The "Cognitive World" (Uexküll's inner world), in which a given organism carries out its specific capacity to determine or control the course of events as they unfold (see above) is replaced in our planning model by the concept "Elaboration of Instructions". After all, planning consists in no small part of coming up with instructions (meaning plans, descriptions, etc.), which is to say devising a course of action in Uexküll's sense. With the aid of these instructions, third parties ought to be in a position to carry out certain actions, such as, for example, building a house or redeveloping a neighborhood.

In order to adapt Uexküll's model to the demands of planning, we will not only "translate" the "Effectual World" into another concept—as we have done with all of the other concepts so far—but we will also divide it in two. This is because in the transition from the "Planning World" to the "Life World" we are dealing with several parts of a complex task. The model therefore makes a distinction between "Communication about Behavior" and "Interventions" (see Figure 2.2). In more concrete terms: in the present context, planning must be understood as a social activity, meaning that in the transition from the "Planning World" to the "Life World" primarily concerned not with the direct and immediate implementation of the instructions developed in the "Planning World" (plans, descriptions, etc.). Instead, the most pressing task is usually to bring what was developed in the "Planning World" into accord with the concerns of the "Life World's" inhabitants. That is because doing so often entails a modification of, for example, the instructions that had been developed in the "Planning World". Hence, we have designated this first portion of the transition between the "Planning World" on the one side and the "Life World" on the other side "Communication about Behavior".

The second portion of this transition we have designated "Interventions" because it entails making material changes in the real world, which take place exclusively within the "Life World" (see Figure 2.2).

As is the case with the "Effectual World", the "Setting" has likewise been split in two. This is done so as to differentiate between a situation *before* and *after* the "Intervention" of planners. For the former situation, in which planners have yet to intervene, we maintain the concept "Setting". It is important to keep in mind that here we are referring only to that portion of the "Life World" to which an "Agent" has the necessary access to carry out observations and then "Interventions". Note that this constitutes a major departure from the "first generation" of planning theories insofar as those theories assume that it is possible to apprehend every feature of a planning situation in its totality (see above). The second portion of the (Uexküll's-)"Setting" we shall refer to as "Outcomes". By this we mean all that an "Intervention" has modified some portion of (planning model's-)"Setting".

In short, we have made a distinction between three aspects of the "Life World": "Interventions" intervene with certain portions of the "Setting" that obtains in the "Life World" and bring about (more or less sought after) "Outcomes".

On the Treatment of Particular Sections of the Model

In the figure above, we have made a distinction between all of the model's constituent elements; however, this does not mean that any section of the model can ever be separated from the rest. The understanding one has gained of a certain state of affairs serves as the basis and conceptual framework for the instructions to be elaborated. The instructions, in turn, influence the manner in which one communicates about the behavior, and so on. Notice, therefore, how arrows tie all of the concepts in Figure 2.2 together in such a way as to give the whole model the form of a circle. However, these arrows only indicate the main path of each element in the whole process. It is crucial to note that they do not necessarily represent the actual order in which every step of a concrete planning assignment must be carried out, since in the majority of all cases planning is a process that includes a certain degree of back-and-forth movement. In addition, many of the individual concepts depicted in Figure 2.2

(such as "Comprehension of the Situation", for example) hide the fact that they may themselves have the form of a circle.

Avoiding the Concept "Rational"

Some readers may have noticed that the above description of our model avoids mention of the concept "rational", as well as any related concept. These days, no planning model should make implicit use—either directly or indirectly—of any idealistic notions of rationality. There are many reasons for this, not the least of which is the fact that philosophy has idealized the concept of rationality to such a degree that it no longer applies to anything we might call a "normal" person (see Cherniak 1992). Thus, for example, when Lenk and Spinner analyzed the term "rational", they came across no less than twenty-two variations in its meaning. The only property common to all these variations is the fact that "the expression 'rational' deals *in some way or another* with strategies for solving problems systematically" (Lenk and Spinner 1989, 1; italics in original). It is no surprise, then, that the concept "rational" has fallen out of favor among planners.

It might be argued that, being based on basic principles of biology or, even, anthropology, the proposed model is capable of describing actions that are not usually thought of as being connected to "rational" planning. This is because the framework of the current planning model does not prescribe how well thought-out or well-founded every single stage of the planning process should be, nor whether the planners ought to proceed carefully or without much deliberation. The scope of the model can therefore be characterized as follows: it does not preclude descriptions of circumstances that include elements such as chaos, confusion, helplessness, and operational inefficiency (see Heidemann 1995). For example, the "Setting" in the "Life World" often develops into chaos. Of course, in this case it remains to be seen whether the "Setting" is indeed itself chaotic or if it is only we who perceive it as being chaotic. In these types of situations, errors of perception, misunderstandings, and confusions are common, and a well-founded "Comprehension of the Situation" is therefore either difficult to achieve, or, if it has been achieved, it is often lost. This loss is usually accompanied by a feeling of insecurity. From this, helplessness follows. Any and all subsequent efforts at "Intervention" are an uncoordinated, blind running around, or the work of hectic and inefficient busybodies. Everyday experience is enough to prove that this happens not infrequently.

Independent of such extremes there are of course also intermediate cases, in which individual sections of the planning model are reduced to an absolute minimum and at times nearly blend together with other sections of the model. What has come to be called "rule-governed planning" serves as a good example of this phenomenon (for further details, see Reason 1990).[9] In this kind of planning, general recommendations such as "problems of the type X are best solved by Z!" are used. Two examples: "The problem of traffic jams that occur whenever interstate highways pass through a town or city is solved by bypassing those towns or cities." "Park-and-ride lots at

---

9   The expression "rule" is not used in the same sense by Reason as it is in Chapter 3.12 below; for further details, see below.

railway stations are best placed far from a city, because motorists tend not to change modes of transportation once they can actually see their destination (i.e. the city)." Rule-governed planning seeks to reduce the understanding of a given state of affairs to a minimal amount of key data in order to subsequently make the decision of how to proceed simply by applying a general rule of thumb. Hence, in this case the sections of our model called "Comprehension of the Situation" and "Elaboration of Instruction" are closely tied to each other at the same time as they are robbed of nearly all their complexity. Rule-governed planning is relatively widespread and, in terms of the amount of effort it requires planners to invest in a given project, quite cost-effective. Unfortunately, it often leads to mistakes, such as when the given recommendation for how one should move forward does not prove to be a good fit for the problem.

Lindblom's "Science of Muddling Through"

Even Lindblom's ideas (1959) can be assimilated to the proposed planning model; after all, he remains the "chief spokesperson" (Hudson 1979, 389) for criticisms of the "first generation" of planning. To summarize his "Science of Muddling Through": Activity is triggered by unsatisfactory situations. Planning consists of focusing exclusively on interventions that the planner is *de facto* capable of controlling, understanding, and planning out. If it turns out that he is not really capable of doing these things, then at least the above represent his first and most immediate actions. His activities are not meant to reach higher-order or more long-term goals. Rather, they are meant simply to "heal" some of the more pressing aspects of a larger problem. Only a few of all the alternative courses of action and a limited number of all the possible outcomes are thereby taken into consideration. No real attempt is made to define a problem, or even a concrete goal, completely. Accordingly, the planner is not really expected to discover an ultimate solution that will stand the test of time. Rather, it is expected that each problem will have to be re-defined continually as time goes by. Since planners often neglect substantial criteria of the problem and important consequences of the proposed solution, the planner's solution often gives rise to new problems later on down the line. As such, we can view a given problem as constantly being in flux. Therefore, planners must continually tackle an old problem's latest manifestation with novel solutions. This strategy can best be described as "muddling through" (see Lindblom 1959).

Lindblom puts the bulk of his emphasis on the following three aspects:

- Plans are put into action with the utmost expediency. Planners forgo protracted analyses that are meant to engender a deep "Comprehension of the Situation".
- Lindblom emphasizes focusing on the "Interventions" that planners intend to carry out, and he voices his preference for those "Interventions" that can be put to the test without much hesitation.
- As plans are carried out with all deliberate speed, a planner is also capable of correcting mistakes quickly if his "Interventions" (i.e. his planning measures) should fail to bring about the desired "Outcomes" (see also Popper 1987).

All of these aspects can be mapped onto our model of the "third generation".[10]

Excursus

Before we can explain the exact meaning of all those concepts that are important in the proposed planning model—concepts such as "Planning World", "Life World", "Comprehension of the Situation" and "Elaboration of Instructions"—in the next section of this book, it should be noted that our "third generation" planning model is compatible with two important theories in the psychology of perception and intelligence. Both theories have played an important role in the development of our planning model.[11]

Neisser's Perceptual Cycle

Our planning model mirrors Neisser's perceptual cycle (1979). The concept "schema" is of central importance here. "A schema is that portion of the entire perceptual cycle that exists within the perceiver, which can be modified by experience, ... and which is specific to that which is perceived." (Neisser 1979, 50) Schemata are conceptual frameworks or structures of knowledge that contain presuppositions or expectations about specific objects, people, or situations (see Zimbardo 1992, 623). According to Neisser, the perceptual cycle consists of three phases: Schemata change within the human "Cognitive Apparatus" based on the information available about everyday objects; in our planning model, this phase is analogous to the structures of knowledge, presuppositions, etc. of "Agents" in the "Planning World".

Schemata govern the process of exploration according to which we select the observed objects (in the planning model, these are portions of the "Setting" as components of the "Life World"), which is to say the available information (our

10 Some of Lindblom's additional arguments might clarify this point further: "Making policy is at best a very rough process. Neither social scientists nor politicians, nor public administrators yet know enough about the social world to avoid repeated error in predicting the consequences of policy moves. A wise policy-maker consequently expects that his policies will achieve only part of what he hopes and at the same time will produce unanticipated consequences he would have preferred to avoid. If he proceeds through a *succession* of incremental changes, he avoids serious lasting mistakes in several ways. In the first place, past sequences of policy steps have given him knowledge about the probable consequences of further similar steps. Second, he need not attempt big jumps toward his goals that would require predictions beyond his or anyone else's knowledge, because he never expects his policy to be a final resolution to the problem. His decision is only one step, one that if successful can quickly be followed by another. Third, he is in effect able to test his previous predictions as he moves on to each further step. Lastly, he often can remedy a past error fairly quickly—more quickly than if policy proceeded through more distinct steps widely spaced in time." (Lindblom 1959/1995, 44).

11 Moreover, the present planning model is compatible with other theories as well, such as the control loop in cybernetics (see Wiener 1948/1968) or classical action theory, the so-called TOTE-Unit of Miller, Galanter and Pribam (1960/1991).

model's "Comprehension of the Situation"). These objects (and therefore also the available information), on the other hand, also modify the schema (which correspond to the structures of knowledge and presuppositions of the "Planning World's" "Agents"). The modified schema again governs the exploration while the modified exploration once again selects new objects ("Settings" that obtain in the model's "Life World"), which is to say available information (the model's "Comprehension of the Situation"), etc.

This brief description will—hopefully—suffice to illustrate how the proposed planning model is built on the basic principles of human perception.

Piaget's Theory on Human Thought and Intelligence

Piaget's work continues to dominate the discussion of the information-processing and adaptive function of human intelligence (see, for example, Piaget 1974 and 1976). He attempts to derive the adaptive mechanism of human intelligence from lower-level, unconscious biological balancing processes. In his work, he attributes to intelligence as its primary function the reciprocal adaptation of the individual to its environment as well as the active adjustment of information about that environment to the needs of the individual. More specifically, according to Piaget, two elementary processes participate in the cognitive development of humans: assimilation and accommodation. Both offer a way for the individual to adapt to his environment. This is to say that cognitive structures are two things at once, namely the product and the prerequisite of individual adaptation. During the adaptive process, the information that an individual takes in ("Comprehension of the Situation" in our planning model) is modified in such a way that it can be incorporated in the extant structures of knowledge. However, in order to fit with the information, or in order not to contradict other extant structures of knowledge, these structures of knowledge (which are the structure of knowledge and the presuppositions of the "Planning World's" "Agents") also undergo modification during the adaptive process. Both forms of adaptation are instances of a more general, underlying developmental principle, namely the equilibrium-principle (see Zimbardo 1992, 65f). According to Piaget, each adaptation is triggered through some kind of a "perturbation", meaning, by a subject's realization that something is out of order, does not function properly, or just seems generally out of the ordinary (see Glaserfeld 1997, 157).

Here, too, we can discern a relationship to our planning model.

So much, then, for this excursus. Let us return to the task at hand, which is to elucidate the constituent parts that make up our planning model.

Elucidation of the Planning Model's Constituent Parts

This section attempts to clarify the concept of a "Planning World" and a "Life World", as well as the different points in the circle described above (see especially Heidemann 1995). At this point, a note of caution is in order: given the scope of this book, the description below will, of necessity, be sketchy and incomplete and will usually only treat one or two aspects of each issue. After all, nearly every individual portion of the proposed planning model offers more than enough substance to fill several books.[12]

Note that we are dealing with two "worlds" in this planning model: the "Planning World" is embedded within the "Life World" and both are related to each other reciprocally (see Figure 2.3). Moreover, we are dealing with a circle that spans both of these two "worlds": a specific "Comprehension of the Situation" forms the basis for the "Elaboration of Instructions". These instructions, in turn, are the basis for the "Communication about Behavior". The result of this discussion governs whatever "Interventions" a planner chooses with which to intervene in a specific portion of the "Setting" in the hope of bringing about certain "Outcomes". The interpretation of these "Outcomes" then leads to a new "Comprehension of the Situation" and so forth.

Planning World

The "Planning World" is the domain in which plans or instructions are developed. As a rule, several planners tend to work together and as part of a larger organization, or "Institution" (keyword: "Institution"; note: those expressions singled out as "keywords" in this and the following sections refer to the terms reproduced in Figure 2.3). The "Planning World" provides the organizational and theoretical background (methods, concepts, theories, worldviews, etc.) for the sections called "Comprehension of the Situation", "Elaboration of Instructions", and "Communication about Behavior".

Key questions here include the following: What are all the different kinds of "Approaches" that exist in the "Planning World", and for which tasks is each one best suited? (Keyword: "Approach") Which methods and procedures do planners actually use, what concepts do they adopt, and what are their worldviews? After all, every "Approach" in planning has its own methodological and theoretical content, which sets the tone for how to proceed in each of the three sections mentioned above.

For example, the following "Approaches" have found use in modern urban planning, either alone or in conjunction with one another: some see the job of urban planning to consist of shaping urban spaces in the most aesthetically pleasing way possible. And/or a planner might proceed by making use of so-called guiding principles of urban development. Concrete options include the separation of functions versus integration of functions, maximum green space versus higher population density (urbanity), cities of short distances, etc. And/or a planner might avail himself of the component-and-relations model developed in systems-theory, such as in the noise-calculations of traffic planning or in calculating the spread of pollution in the air and the ground. Or one might use the system-environment paradigm when planning (see Heidemann 1995 and above), etc.

---

12 The section that will follow only describes the most salient features of the "third generation" in planning theory. Each individual aspect will therefore require further differentiation. Moreover, the themes that will be discussed below have been the subject of protracted debates in other specialized fields (economics, (organizational) sociology, political science, etc.), debates to which we can refer only briefly here.

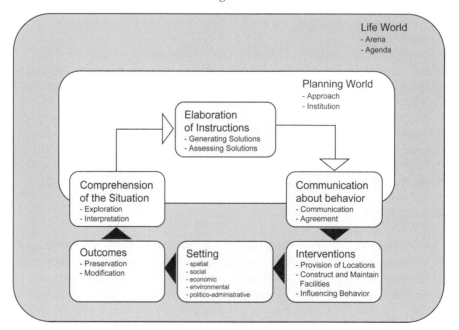

Figure 2.3   Basic scheme for planning with supplementary keywords

Source: Heidemann 1992, 95 (modified).

As different as these various "Approaches" are, the results or solutions that are sought after in each individual circumstance are equally diverse. This means that the "Approach" an "Agent" chooses goes a long way towards determining the results of a particular planning situation. With this in mind, it is noteworthy that there are few comparative descriptions of planning "Approaches" that exist in planning theory (see, for example, Koschitz 1993).

The fact that the particular way each planner happens to see the world plays a crucial role is made clear by the example of traffic planning. Some planners work to streamline traffic by way of particular measures whose goal is to control the way traffic flows. Others, however, try to figure out how to eliminate traffic altogether. In most cases, different political foundations, which in effect are different ways of seeing the world, underlie decisions such as these. In case of the former, it is usually "Cargoism" (see Catton 1980), in case of the latter, it is not infrequently "environmentalism". In the final analysis, then, how we formulate questions in planning depends on how we happen to see the world, and it is these ways of seeing the world that form the basis for how we understand specific problems as well as the values that our proposed solutions to those problems attempts to uphold.

Life World

The "Life World" contains everything that surrounds the "Planning World". The stages "Intervention", "Setting" and "Outcomes" of the planning model are a part of the "Life World" (see below as well as Figure 2.3).

A core feature in this context is the pair of topics "Agenda" and "Arena". The "Agenda" is the given catalogue of political issues available for discussion and debate, which can initiate planning processes (keyword: "Agenda"). Neither the issues up for discussion and debate, nor the occasion for initiating a planning process that results from these discussions and debates are ever purely "objective." Moreover, a lack of funds, time, interest, etc., usually precludes us from simultaneously working on every problem that, in theory, requires planning. Instead, it is usually the case that only a few select problems receive much attention, and correspondingly few "Agents" are willing to tackle those problems. This causes certain issues in planning to have a sort of "career" (see Luhmann 1979), interest in them fluctuates, and they "drive themselves to death". It is possible then to produce interest in issues—like setting an "Agenda"—such as, for example, "Stuttgart 21" or the "green belt" in Frankfurt am Main. Likewise, it is also possible to deflect attention away from certain issues, and thereby keep them from the public spotlight, such as we often see happen just prior to an election.

It is impossible to understand a given "Agenda" without also examining the correlated "Agents"—which act alone or in tandem with one another—as well as their interests and the course of action that are open to them. We will refer to the set of all these "Agents", taken together, as an "Arena" (keyword: "Arena"). This includes citizens, the authorities, firms, interest groups, planners, etc. The reasons why a given issue on the "Agenda" is treated and resolved in one way rather than another usually becomes evident only after having examined the given "Agenda" and its related "Arena" as closely as possible. Equally important is for all the economic, political, social, and ecological, circumstances of a planning task to mirror each other in the "Arena"-"Agenda" constellation.

To summarize: as a rule, what happens during a specific planning process is intelligible only when each individual step of the planning process is examined ("Comprehension of the Situation", "Elaboration of Instructions", etc.; see below). In addition, it is crucial to examine the "Approaches" used in the "Planning World", both within and beyond the organizational limits of the "Agenda" and "Arena" of the "Life World", which themselves provide the background for and thus govern what happens in the "Planning World".

Comprehension of the Situation

By "Comprehension of the Situation" we mean the act of developing a description of a problem in need of planning such that this description represents the task of the planner as accurately as possible. This usually results from an interplay between, on the one hand, empirical investigation as "Exploration" of a given state of affairs and, on the other hand, the "Interpretation" and assessment of the "Exploration's" results (keywords: "Exploration" and "Interpretation"). Since we have no direct access to the

"Life World" that is not filtered through our "Cognitive Apparatus" and thus cannot perceive the "Setting" as such, the "Exploration" as well as the "Interpretation" must be carried out by an "Agent", and the resultant procedure will therefore be governed by that "Agent's" subjective thought processes (see also Chapter 3.1).

Developing a "Comprehension of the Situation" relates directly to the point of contact between the "Life World" and the "Planning World." Whomsoever should deal with the transition from "Life World" to "Planning World"—understanding that, for planners, it is impossible to avoid dealing with this transition—must of necessity concern themselves with certain more or less central questions of science that must be answered at this point of contact: What is the "Life World"? What is it made of? What thought-processes are used in planning? How are these processes shaped? Why are they shaped just so and not otherwise? How well-founded or "secure" are our cognitive models? What forms the basis of this "security"? etc. (see Vollmer 1988, 3). This is to say that, on a more abstract level, semantics (the theory of meaning and truth), as well as epistemology (the theory of knowledge) and ethics (the theory of values and proper actions) play a role here (as they do in other sections of the planning model). In our planning model we assume that at least those aspects from the above list of topics that are relevant for planners are part of their background knowledge, taking into consideration the fact that as a rule they must first be translated into their own specialized discourse.

Several aspects of the philosophical issues surrounding the "Comprehension of a Situation" will be discussed in further detail in Part II of this book.

Elaboration of Instructions

This section of the planning model concerns the elaboration of plans or descriptions; in what follows we will usually refer to them as "Instructions". These "Instructions" detail all that must be done in order to bring about a desired effect. It is important to take uncertainties, gaps in one's knowledge, and risks into consideration when developing these "Instructions". Simple examples of these "Instructions" include ordinary plans such as development plans for architects, work plans (e.g. floor plans, diagrams, cross sections), and reinforcement plans as "Instructions" for a construction crew. "Instructions" are therefore descriptions of "Interventions" that ensure those "Interventions" are successful in bringing about the desired "Outcomes" (see Heidemann 1995).

When we develop these "Instructions", what we are most often really doing is engaging in an interplay that includes, on the one hand, "Generating (various) Solutions" and, on the other hand, "Assessing (those various) Solutions" (keywords: "Generating and Assessing Solutions"). Examples of this include the following: What are the available locations for, e.g., a new concert hall or garbage incinerator and which one of these locations is the most "well-suited" to the task at hand? Which variants of an architectural design are the "best"? Hence, when we "Generate Solutions" we are looking for different possibilities of how to solve a problem we may have come across in planning. When we "Assess Solutions", we (almost always) settle on one of these possibilities and discard the rest.

The kinds of solutions that present themselves depend, of course, on how the problem has been defined or, to put it differently, how we "Comprehend the Situation". Aside from this, we are mostly dealing with a process of coming up with novel ideas and solutions. This is to say that a planner's creativity plays a role here—the techniques he uses to maximize his creativity (see, for example, Schlicksupp 1992)—as well as his capacity to avoid blocks on his creativity. Over the course of the assessment and subsequent selection of a "suitable" solution, the strategy by which to carry out this assessment occupies the bulk of our attention. The spectrum of strategies ranges from the intuitive to the argumentative all the way to the so-called formal methods of assessment, each with its attendant strengths and weaknesses (see, for example, Eekhoff, Heidemann and Strassert 1981).

Communication about Behavior

Once a set of "Instructions" has been elaborated, it is important to come to an understanding with those who will be effected by the course of actions that follow (this usually means those who will have to participate in that course of actions). This often entails a modification of the "Instructions" (keyword: "Communication"). Here, the professional content of the "Instructions" plays a central role. Added to this are negotiation strategies, techniques of communication, some aspects of group dynamics, the problem of using and abusing power (see, for example, Flyvbjerg 1998), and the diverse forms of participation (hearings, advisory boards, ombudsmen, planning cells, etc.).

The result of this consultation ought to be an "Agreement" that determines who will do what when where and how (keyword: "Agreement"). These "Agreements" can take on various forms: from a tacit consent all the way to a formal resolution, such as, for example, with a construction, land-use, or regional plan. Other kinds of "Agreements" include contracts, such as those used by craftsmen and construction companies. The contract we sometimes find in urban planning, in which different parties are obligated to carry out certain duties, can also be counted among these.

It is only with these "Agreements", which mark out the borders of the "Planning World", that the "Instructions", which is to say plans, become mutually binding for all the participants.

Sections of the Life World: Interventions, Setting, and Outcomes

Once the "Instructions" have been created and an "Agreement" has been reached with the participants, the subsequent actions take place in the "Life World", whereby we distinguish three sections: "Interventions" intervene in the actual "Setting" and produce certain "Outcomes".

Interventions

The keyword "Intervention" refers to all that is actually done on the basis of the "Instructions" that have been previously worked out.

This includes answering the question of how one ought to implement the "Interventions" that have been planned out (planning a deadline, planning of operation sequences, planning the use of resources, etc.). This can be done by, for example, applying the techniques of project management (see Sommer 1998).

The following is a central question in this context: what are the actual "Interventions" that planners advocate as a result of their planning? Normally, there are two ways to intervene in planning: first, there is the possibility of setting aside a location for certain uses (keyword: "Provision of Locations"); examples are development plans, land-use plans, regional plans, etc. Plans such as these determine which areas to designate as residential or industrial and where to place schools, green spaces, and highways. The second method of "Intervention" consists of constructing and maintaining facilities (in the broadest possible sense) in these locations (more precisely: to plan out the construction and maintenance of these facilities; keyword: "Construct and Maintain Facilities"); examples of such facilities include buildings— the classic field of architects—and streets, parks, etc. Traditionally, these have been the two methods of "Intervention" available to planners. This view of the profession is also mirrored in some of the classic subjects in planning education, such as Christaller and Lösch's Central Place Theory (sometimes also called Location Theory, see Blotevogel 1995, 1117 ff), or—in the case of Architecture—Building Studies. However, this view of planning poses some problems; namely, it often leads to plans that do not sufficiently take people's actual behavior—that which is supposed to transpire in or on these facilities (buildings, streets, etc.)—into consideration. Both Central Place Theory and Building Studies assume a typical or "average" pattern of behavior. These are sufficiently accurate to prove useful in many cases, to be sure. However, they lead to complications whenever the existing way of using facilities changes, or when new ones develop.

In our judgment, the job of planners has at least three parts, rather than just the two described above (viz. the "Provision of Locations" and "Construct and Maintain Facilities" in these locations). In principle there is also the possibility—which makes up the third way to intervene in planning—of influencing the behavior of those who make use of the aforementioned facilities by, for example, instituting new rules or ordinances that legislate the manner in which facilities may be used. This is to say that planners need not change anything about a built structure *per se*, but only the way in which this structure is used (keyword: "Influencing Behavior"). Many real-world examples come to mind: there are not only streets, but also rules of the road; residential communities, car sharing, car-pooling, toll roads, management of commercial traffic, etc. These are all attempts to influence and regulate the use of certain facilities. Moreover, the use of a facility can be influenced by means that are not only contractual and legal, such as in car sharing and traffic laws, but also educational, i.e., when factual information is used to elucidate that certain behaviors are prudent whereas others are not. Examples of this second method of "Influencing Behavior" through education include systems to direct traffic or information-systems that help people navigate public transportation (train, bus, etc.). Over and above the effort simply to make information available to the public, it is also possible to actively invite, encourage, and motivate people to act in a certain way: to use public transportation instead of a car, for example; to separate some garbage as recyclable instead of throwing it all on a landfill; to ventilate a house in an energy efficient manner; to use less electricity; etc. Of course, it is also possible to use financial incentives to influence behavior (promotional programs, tax rebates, etc.).

All of this is to say that "Influencing Behavior" is one of the tasks of planning; in actual practice, planners come across problems whose solutions require that the planners influence people's behavior anyhow: examples include the domain of traffic, energy, and environment.

Setting

By "Setting" we mean the totality of all the things that exists in the "Life World", in all their spatial and temporal particularity, which we hope to somehow modify or preserve through planning. More concretely, we are here talking about that portion of the "Life World" to which the "Agents" have access for the purposes of observation and action. We are, finally, dealing with all things that are valuable as well as all things that are deplorable in our world—more precisely, everything amongst these that planning can somehow influence. In this context, it is important to note that it is up to individual planners (in the process of achieving a "Comprehension of the Situation", which is to say, by choosing a specific "Approach" to planning; see above) how much of the "Life World" he will take into account when working out his task. He must decide if he will include every relevant spatial, social, economic, ecological, and political (i.e. administrative) aspect of the "Life World" in the process of working out a plan (keywords: "Spatial, Social, Economic, Environmental, and Politico-Administrative"). A planner's choice in these matters helps him determine the adequacy of the developed plan relative to the task at hand.

Outcomes

The term "Outcome" serves as shorthand for everything that has established itself as a result of carrying out the planned "Interventions". Note that we do not here make a distinction between "Outcomes" that comply with the originally intended results of the planned "Interventions" and those that do not. We do speak of a successful plan in the former instance. After all, planning is almost always a matter of improving a disadvantageous situation (keyword: "Modification") or maintaining an advantageous one (keyword: "Preservation").

The central question is always if we can test whether or not a specific plan was indeed successful. As a matter of fact, it is rarely possible to ascertain whether a specific measure in planning has been successful or if it was a failure. Hence, we must make a distinction between at least two basic problems: the first question is if "success" even exists as such, and how long it takes for "success" to take hold. It is especially difficult to make this judgment for so-called prophylactic measures. That's because these "Interventions" aren't meant to eliminate some disadvantageous or harmful situation; rather, they are meant to avoid an unwanted situation before it even arises. When, for example, can we say that an environment suitable for raising children has been created successfully? After one year? After five, ten … fifty years? These kinds of preventative measures are intentionally designed such that, if they are successful, *nothing* (negative) happens.

The second question is if the observed result was really brought about by the planned "Interventions". Luck, chance, or unknown circumstances could just as well have caused the observed results. Strictly speaking, it is only possible to ascertain the effectiveness of a given "Intervention" if there is a second situation to which we can compare the first such that the only thing that distinguishes the first situation from the second is the fact that the specified "Intervention" has been carried out in the first instance and not in the second. If it should then turn out that nothing changed in the second situation whereas in the first, planned-for situation the desired "Outcome" materialized, then it would be safe to say that the "Outcome" was caused by the planned "Intervention". This, however, is purely utopian speculation; planners never have access to scientific controls of this kind. This, then, is why we have a second difficulty in evaluating the results of planning.

Both of the above points make it clear that suitable strategies must be applied (for example ex-ante-analysis, ex-post-analysis, etc.; see Campbell and Stanley 1966 or Patton and Sawicki 1993), to assess the effectiveness of a given "Intervention".

Cognitive Traps of Individual Sections of the Planning Model

There is rarely just one way to work through any individual sections of the planning model ("Comprehension of the Situation", "Elaboration of Instructions", etc.). Rather, planners are often presented with a whole spectrum of possibilities. The two outermost points of this spectrum represent the most "scientific" and the most "spontaneous" strategies. If only because planners have limited amounts of time available to them, their actual strategies often fluctuate, moving back and forth somewhere between these two extremes. More often than not, a truly diligent, careful, and well-founded method of working through every single aspect of a problem in planning is impossible. If for no other reason than this, guidelines that make us aware of typical errors that often come up when working through an individual section of the planning model are worthy of our attention. These guidelines include, for example, some empirical findings in cognitive psychology, which show that our planning, just as our thinking, is based on a whole series of unconscious and inherent cognitive tendencies (often traps). These cognitive traps are mirrored in the errors we make when planning (see Schönwandt 1986 or Dörner 1989). Table 2.1 therefore lists, in summary form, some of these cognitive traps and specifies during which section of the planning model they usually emerge. This, it is hoped, also has the additional virtue of clarifying how further distinctions made within each individual section of the planning model can help us understand and describe the actual processes that planners go through when planning.

Classifying the Seven Models of Planning from Chapter One

Even if it has already become apparent how the seven models of planning described at the beginning of this book fit into the "third generation" theory of planning, we will quickly sketch out such a classification below. A word of caution, though: the outline below will confine itself only to a few of the most relevant aspects.

At first sight, the *rational model of planning* (Chapter 1) seems similar to the circuit of the "third generation", which runs from the section called "Comprehension of the Situation" to "Outcomes". In fact, there are important differences though, some of which are reproduced in the table below (see Table 2.2).

Moreover, the "third generation" theory of planning incorporates two additional levels of observation absent from the rational model: first, there is the "Planning World" with its organizational boundary conditions ("Institutions") as well as the

Table 2.1   Cognitive traps in planning

Comprehension of the situation
We have a tendency to:

- ignore problems and chiefly take actions only as a response to obvious and undeniable difficulties,
- overlook the bulk of potentially useful information,
- search primarily for information that we would like to find, and to ignore information that contradicts our own assumptions,
- carry out a superficial analysis of the situation, form an opinion on the basis of only a few key facts, and use that information as a basis from which to make erroneous projections,
- assume that trends will continue in a more or less linear fashion,
- regard some information as true, though it is unquestionably false, when pressed for time,
- make unrealistic estimates of how much time a certain process will require.

Elaboration of instructions
We have a tendency to:

- plan using rules of thumb rather than conduct careful analyses of the problem,
- designate various solutions as either good or bad before actually having understood them,
- implement the first half-way acceptable solution to a problem, rather than systematically looking for further solutions,
- look for further solutions, if the initial choice for a solution has failed, only in the "near vicinity" of this first, initial choice,
- not look for promising alternatives when it becomes obvious that a solution is not suitable, but rather continue to invest in actions that are already known to be ineffective (to entrap oneself).

Communication about behavior
We have a tendency to:

- confuse agreement on an issue by members of a group with truth,
- underestimate the risks of a plan when acting as part of a group ("collective blindness"),
- act in a way that ensures our social relations with other people are not jeopardized, rather than to act and make decisions based on the facts.

Interventions

We have a tendency to:

- overestimate the effectiveness of a planned intervention (the illusion of control),
- overlook after- and side-effects of a measure in planning.

Setting

[The Setting cannot be perceived directly, but only through a "Comprehension of the Situation"; see above.]

Outcomes

We have a tendency to:

- say "I knew it all along" after the outcomes of a plan have manifested themselves, and thereby unconsciously cover up how much we did not know;
- we find it difficult to assess what caused certain outcomes of a plan objectively, which is to say that we have difficulty assigning responsibility for success or failure appropriately (failures are often misinterpreted as a success; skill and

"Approaches", all of which make up the conceptual backdrop for processing a task in planning. Second, there is also the "Life World", into which the "Planning World" is embedded. Both worlds stand in a reciprocal relation to each other.

The *advocacy model*, the *model of equity planning*, and the *model of social learning and communicative action* focus on "Communication about Behavior" with a more or less pronounced interest in the political "Agenda" and "Arena" of the "Life World" by concentrating on communicative action, which is to say by pushing those who will be affected by the plans into the foreground of our attention.

Planners that use the *radical model of planning* construct their own "Planning World", with their own organizational boundary conditions and their own conceptual "Approaches" (methods, concepts, theories, and worldviews), by working in opposition to state-sponsored organizations for planning and by turning their backs on and working outside of the traditional methods and official procedures of planning. In the eyes of official planners who work for state-sponsored organizations, these radical planners are often seen as agents in the "Life World" that attempt to implement certain aspects of their "Agenda".

In the *(neo)Marxist model of planning*, the (neo)Marxist worldview forms the focus and starting point of any deliberations in planning. This theory/worldview is a portion of the "Approaches" in the "Planning World" and therefore influences how a specific task for planning is worked out.

In the *liberalistic model of planning*, just as in the (neo)Marxist model, a specific theory/worldview makes up a portion of the "Approaches" in the "Planning World" and thus plays a dominant role in the course of actual planning. In this case, it is the ethical theory of so-called Liberalism (for details, see, for example, Mill 1859, Nozick 1974, and Hayek 1976).

Table 2.2   The rational model of planning versus the "third generation" of planning; comparison of some assumptions

Some assumptions that form the basis of the rational model of planning	Some assumptions that form the basis of the "third generation" theory of planning
• information is fully apprehended, (completeness)	• Information is never completely available, the state of a problem can only be understood piecemeal
• Courses of action are "objective"	• Courses of action are subjective; we have no direct access to the "Life World" "as it is". Every act of perception and cognition is theory laden and therefore never "value-neutral"
• Courses of action are "rational"	• Courses of action are influenced by, among other things, cognitive traps
• Formulation of a problem and solving a problem are separate and independent processes	• The individual sections— "Comprehension of the Situation", etc.—are distinguished from one another but cannot be separated
• Solutions ought to be "optimal"	• There are no "optimal" solutions, especially because the parties involved almost always have different preferences, which, moreover, are not always stable over an extended period of time

In individual cases, other classifications are also possible, depending on which facet of the planning model is pushed to the foreground.

Summary

To summarize, the "third generation" planning theory outlined above contains the following components (note that this depends on *all* the aforementioned aspects working together): "Agents" of planning, who tend to act as part of a specific organization, construct a "Planning World" in their respective "Cognitive World" (methods, concepts, theories, worldviews, etc.). This "Planning World" exists in the context of a "Life World", and within it a specified "Agenda" of issues is treated by the agents of an "Arena". A specific manner of exchange always occurs among both "worlds". This exchange happens mainly at the foundation of a functional circle in which the following sections are distinguished from one another, even though they cannot be separated from one another: "Comprehension of the Situation", "Elaboration of Instructions", "Communication about Behavior", "Interventions", "Settings", and "Outcomes". In this process, the outcomes can engender a new "comprehension of the situation", and therefore at times also new planning processes. Particular problems are worked on in each one of these individual sections. Moreover, typical cognitive tendencies, i.e. cognitive traps, appear in these sections.

What happens in a concrete and specific case in planning can therefore only be grasped when the given steps of the planning model are scrutinized ("Comprehension of the Situation", "Elaboration of Instructions", etc.). The "Approaches" (see above) used in the "Planning World", as well as the "Agenda" and "Arena" of the "Life World" (each one of which provide the background for what happens in the "Planning World") must also be scrutinized.

Who can realize such a model? Of what use could it be? A few comments may suffice: for one thing, the model is well suited for making manifest the often extremely limited perspective that develops when treating a planning task. It can also contribute to maintaining a more general overview in a planning process. Moreover, it can help to make the complexity of a planning task manageable and the course of events more transparent, as well as improving the understanding of all the parties involved. Finally, it improves the odds that partial tasks are not only treated successfully by individual people, but that the results of their labor are then re-combined sensibly as well.

The thoughts represented here describe the basic features of a "third generation" in planning theory, though the individual aspects require further differentiation. The sheer number of individual topics that must be taken into consideration may irritate some; however, there is no easier way to achieve a well-founded plan, as well as a well-founded planning theory.

2.1 Aligning the Positions Here Advocated with Actual, Mainstream Approaches in Planning Theory

The planning model described here in Chapter two as well as what follows in Chapter three are a contribution to the further evolution of planning theory. One question remains to be answered: how to align the positions articulated in these chapters with mainstream approaches in planning theory? An answer to this question will be the topic of the brief section that follows.

A major obstacle that needs to be overcome right at the outset is to determine which actual approaches in planning theory are to count as "mainstream" and which do not. After all, there are almost as many schools of thought on the subject as there are forums in which to debate them. For the sake of economy, the discussion that follows therefore confines itself to three central, leading, and prominent positions: collaborative/communicative planning, postmodern planning, and post-positivistic planning (see Allmendinger 2002, see also Harper and Stein 2006).

To simplify the task that lies ahead, it is helpful to describe right away a central assumption that lies at the foundation of all the issues discussed in this book, and hence the present section as well. In the following passage, Min (2001, 327f) provides a good summary of the assumption in question:

> There is no perceptual form of rationality, no universal framework or reference of meaning, no absolute notion of truth, knowledge and scientific law and no set of unchangeable, universal rules, principles, methods or procedures in knowledge acquisition and

justification… There is no everlasting consensus over moral issues, cognitive judgments and aesthetic appreciation… Therefore, there is no absolute certainty in the traditional sense.

These assertions hold true quite generally—they apply to all kinds of knowledge—and, hence, they also pertain to the knowledge used in and generated by planning. In addition to what is expressed by the quoted passage above, the scientific community also recognizes a whole series of generally well-established conclusions that agree with the assumption under discussion here. All of these conclusions tend in the same direction. They include:

- Knowledge is not a bundle of facts, but a collection of ideas about facts.
- We—as human beings—have no direct access to the outside world.
- We are never acquainted with things "in themselves".
- Every description includes aspects of the cognitive world of the describer.
- Every description of facts is theory-laden.
- All knowledge is socially constructed.

The reader will find it helpful to keep these conclusions in mind when reading the text that follows.

Collaborative/Communicative Planning

Given what has been said so far, it follows almost immediately that the first of the above-named approaches to planning—the collaborative/communicative approach—must be, and indeed is, seen as "state of the art" and foundational for the issues covered in this book. This is one reason it was already described in chapter one. It is also why so much has been published about this approach in recent years. (For a discussion on the difference between collaborative and communicative planning see, for example, Tewdwr-Jones and Allmendinger 2002, 208ff.) In somewhat simplified form, the central tenet of the collaborative/communicative approach may be summarized as follows: in principle it is not possible for us humans to understand, describe, or act on anything *without doing so on the basis of some cognitive process and from some cognitive viewpoint: there is no vision from nowhere.* Everything that we see, think, plan, say, and do is influenced by our perceptions and experiences. We see the world as if through a pattern or mould of our own making into which we then try to fit our experiences of that world. Since there are always a great many people involved in planning, it follows that there will always be manifold viewpoints and stances on issues rather than a single, "true" matter of fact.

In planning theory, the realization that all knowledge is situated in this way has resulted in a critique and indeed wholesale abandonment of the old, "rational" planning model. This is in part because—at least implicitly—the "rational" model assumed (see Chapter one above) that a planning situation can be described in terms that are purely "objective" and do not take into account the experiences of individual subjects.

The results of introducing this notion of situated knowledge have been and continue to be pervasive. There is no *single*, objective "view from nowhere", but rather a manifold of different views from somewhere. Each of these represents a different point of view and a different, situated form of knowledge. Hence, if we want to develop solutions to a given planning problem, we cannot avoid acquainting ourselves with all of the different points of view that various people bring to the problem at hand. Clearly, we can only acquaint ourselves with these different points of view via a process of communication. This means there are good reasons for planners to engage in the study of communication.

However, in what follows we will see why this engagement alone, no matter how intense, does not suffice.

The Turn to Content

This book (see here especially chapter three) makes the case for a turn to follow the "communicative turn" discussed immediately above, namely the "turn to content". In order to forestall misinterpretation, let us immediately clarify that by advocating a turn to content we do not thereby claim the issues currently discussed under the heading "communication in planning" are therefore irrelevant or superfluous. Rather, we merely wish to add emphasis to a point that is often overlooked or given insufficient treatment in the discussions on communication: the conceptual formation of content, i.e., knowledge. Differently put: "communication" is important, but no more important than the question of what a given act of communication is *about*. It is true that some advocates of the turn to communication—Habermas is a prominent example here—understood that the content of communication, which is to say knowledge, must form an important part of any discussion about communication. However, the mainstream of the contemporary discussion on communication in planning theory often ignores this aspect of their chosen topic.

Communication versus Content

The shortcomings of the collaborative/communicative approach to planning with which we are concerned can be described as follows: there is a difference between the conceptual contents of a planning task on the one hand and, on the other hand, issues such as "communication" and "discourse", meaning the *historical, social, psychological, etc., conditions that make it possible to process and convey* this content in the first place. These two issues—"content" and "communication"—are often at least implicitly and incorrectly assumed to be identical. Above all else, the word communication here refers to the aforementioned conditions that make it possible to create and process the content, rather than the content itself: In this context, "communication" is similar to "transport". That is because the word "transport" similarly leaves unanswered the question of what it is that is being transported. "Communication" and "discourse" *influence* the processing of conceptual contents, yet they are *not* to be *equated* with the contents themselves.

The current focus of the discussion on communication is also evidenced by the issues on which the communication process in planning usually focuses: Who should be included? How is the process of communication to be organized? How is it to be moderated? According to what rules are decisions to be reached? What role do the power dynamics of the various participants play? *Et cetera.*

Given all of this, an important argument of this book can therefore be summarized as follows (for further details, see also Schönwandt and Jung 2006): current discussions on communication have turned a blind eye to the content of communication (even though, as has already been mentioned above, this shortcoming does not infect the work of everyone who inspired the current discussion).

Content/Knowledge

The preceding almost immediately begs the next question: what do we mean when we talk about "content"? About "knowledge"? Here, we must first distinguish between, on the one hand, knowledge that can be expressed in language and, on the other, tacit knowledge (see Bunge and Mahner 2004, 64). The latter is usually exercised unconsciously, especially by artists and artisans. Since we are here dealing with planning as a process of communication, we will restrict our discussion to that kind of knowledge which can be expressed in language. The reason for this is that tacit (i.e., unarticulated) knowledge cannot be exchanged through acts of communication.

The next question is therefore: what does this expressible knowledge consist of? It always consists of "concepts" that can be tied to "propositions" as well as "guides to action" (see Bunge and Mahner 2004, 37). "Concepts" and "propositions" are therefore the basic building blocks of this kind of knowledge.

An example will help explain the point: The planning assumption of a "city of short distances" includes *concepts* such as "city", "density", "functionally heterogeneous neighborhoods", "traffic", etc. In addition, this planning assumption rests on the following *proposition*, in which the aforementioned concepts are brought into a contextual relationship: "dense and functionally heterogeneous neighborhoods allow for shorter distances and thereby produce less traffic". In turn, this proposition can be used as the basis for a *guide to action* in planning: "the goal of producing less traffic can be achieved by building dense and functionally heterogeneous neighborhoods". (See Chapter Three.)

"Concepts" and "propositions" as basic building blocks for linguistically expressible knowledge are therefore central and pivotal to the present subject. Their prominent and often undervalued significance can be summarized as follows: Concepts and propositions are the bearers of our knowledge and therefore inform our actions in planning. Consequently, they largely determine what is planned and what results emerge from the planning process. A central assumption here is that the more well-founded our concepts and propositions are, the more well-founded the knowledge that underlies our plan will be. Hence, the more likely it will be that the implementation of our plan will also lead to the desired ends. The success of a plan therefore depends on the coherence of the concepts and propositions we use. Together, the concepts and propositions form the foundation on which everything else is built, as it were. To extend the metaphor: if this foundation cannot "carry the necessary weight", then everything else is in "danger of collapse".

This is to say that many failures in the execution of a plan can be explained as failures of knowledge. This is why it is vital to develop whatever concepts and propositions we use carefully and to make sure they apply to the situation at hand, not least because the planning measures that follow tend to have far-reaching consequences on the living conditions of people. For example, some people make a fortune when a plan calls for their pasture or meadow to be converted into a construction site. Others have to endure the effects of air and noise pollution that is produced when their neighborhood is turned into a commercial zone, or a railway or highway are planned and subsequently built near where they live. Those who are affected by various planning measures therefore have a right to expect those measures to rest on solid and durable knowledge rather than empty verbiage or pure speculation.

Objections to the Theory

Someone might object that those working with concepts and propositions are only playing some kind of academic game. Nothing could be further from the truth. None of us (and this includes the most hard-nosed, practicing planner) has a choice of whether or not to use concepts and propositions. After all, planners (and scientists) never work with reality "in itself"—in the Kantian sense. Rather, we only have access to *descriptions* of reality that are based on our perceptive and cognitive processes. This is true even if we are unaware of it, as the fish is unaware of the water that surrounds him on all sides. These descriptions of reality consist of what we have called the basic building blocks: concepts and propositions. All we can do is to choose between relatively nebulous concepts and clearer concepts on the one hand and relatively ill-founded propositions and more well-founded ones on the other (see, for example, Campbell 2003). Nobody can get around concepts and propositions, whether we are aware of it or not.

We admit that this Kantian idea—that we never work with things "in themselves" but rather just descriptions of those things—is not especially easy to grasp. The official advisor of a German Research Foundation project on the use of concepts in planning makes us aware of this when he admits that some colleagues may find this "problem difficult to approach without prior and extensive education".

For these reasons, chapter three is dedicated to giving a detailed treatment to the topic of concepts and propositions.

Postmodern Planning

The second of the mainstream approaches enumerated above is postmodern planning. A start on unpacking this approach can be made by remembering why the advocates of postmodern planning criticize earlier planning models. Sandercock (1998a), for example, provides an excellent example of the critique that postmodernism levels against the "rational" model while simultaneously indicating some ways it might be improved:

> "Postmodern critiques begin with an attack on the very idea of a possible theory of knowledge (or justice or beauty), arguing that the pursuit of such theories rests upon the modernist conception of a transcendent reason, a reason able to separate itself not only from historical time and place but also from the body. "There is no Reason; only reasons" ... The critique of the Enlightenment concept of rationality and its unitary definition of truth

forms the basis of postmodern thought. The postmoderns claim that the Enlightenment privileged rational discourse by identifying it as the sole avenue to truth, and by defining that discourse in terms of its abstraction from social context. Postmodernism rejects this privileging of rational discourse, arguing that there cannot be one such privileged discourse." (Sandercock 1998a, 70)

"... all commit the epistemological sin of assuming one overarching truth." (Sandercock 1998a, 71)

The explanations given at the beginning of this section should serve to make it clear that the arguments of postmodernists are already accommodated in the proposals advocated by this book. This book will explicitly refuse to countenance any notions such as "one overarching truth" or a "sole avenue to truth". On the contrary, in planning we are always and of necessity dealing with a manifold of different points of view that are formed on the basis of individual perceptions and experiences and that influence what we see, think, plan, say, and do (see above). A central concern of planning is therefore to integrate, as much as possible, all of these various points of view. (Doing so is anything but easy, not least because of the ethical conflicts that are often involved.) This is one of the fundamental positions for which the present volume argues. Sandercock (1998a) has a perceptive diagnosis of the main task that confronts all planners: "The overall goal is ... to generate a political process, which may involve plans, policies, programmes." (Sandercock 1998a, 7) ... "In view of this, we have to somehow find ways of collaborating with each other, without retreating to outworn political models, to assert our needs and desires and social ideals in the face of ... global economic rationalization." (Sandercock 1998a, 7)

In this context, we would be remiss not to mention, however briefly, another foundational methodological point. The postmodern critique enters the discussion at hand largely because earlier planners had been working under a rather naïve view of scientific practice—a view that, for example, assumed there to be such a thing as "one overarching truth" or a "sole avenue to truth". In this vein, Gabriel (2004, 306) makes the following observation: "The claims made by advocates of postmodernism ... only become plausible against the backdrop of a dogmatic metaphysics and a naïve philosophy of science that results from a misunderstanding of scientific research as objective and value-neutral."

As the planning model this book defends explicitly lacks such a naïve view of scientific practice, it fully meets the demands of the postmodern critique.

Post-Positivistic Planning

What is true of the postmodern approach to planning generally holds for the third mainstream approach mentioned at the beginning of this section—post-positivistic planning—as well. As with the other two approaches, those aspects of planning that are deemed important by the post-positivistic approach are taken into account by the arguments presented in this book.

One of the most central tenets of the post-positivistic approach to planning is well illustrated by the following passage from Koppenjan and Klijn (2004, 37): Previous models of planning ignore "... the social aspects of knowledge production and thus the fact that research departs from a certain definition of the problem. This insight is characteristic of a post-positivistic look at science and society [and planning] ... Knowledge is developed in an interaction process between knowledge producers who make subjective decisions partially inspired by societal influences. Research is not separate from other societal subsystems, it is entwined with them." (For a depiction of the post-positivistic approach in greater detail, see, for example, Allmendinger and Tewdwr-Jones 2002). Clearly, then, the post-positivistic approach rejects any "sole avenue to truth", just as the postmodern approach did. In this respect the two are substantially similar.

A substantial difference between the two approaches in the context of planning is as follows: the post-positivistic approach makes an additional critical point about discussions in planning theory that is left out by postmodern arguments. To wit: it points out that planning-theoretic discussions often and wrongly presume that the collaborative/communicative approach is not just the currently dominant approach to planning, but that, in addition, it has the status of a paradigm (*sensu* Kuhn 1962/1981; see, for example, Innes 1995). In response to this mistaken view, Tewdwr-Jones and Allmendinger (2002, 214) formulate the following counter-argument: "[C]ommunicative and collaborative planning is ... a 'world view' based on a participatory perspective of democracy and either a suspicion of or a more balanced attempt to situate free-market economies: Its context [is] planning as a democratic enterprise aimed at promoting social justice and environmental sustainability" (Healey, 1997:233)."

Tewdwr-Jones and Allmendinger (2002) make an important point here: the collaborative/communicative approach to planning is enmeshed in a larger, ethical worldview, namely the social justice championed by thinkers like Rousseau, Marx, or Rawls (see, for example, Davy 1997, 267). However, this is only one of many possible worldviews. After all, there are others such as the elitist or libertarian justice of Nietzsche, Smith, or Hayek (see Davy 1997 or Harper and Stein 2006). The collaborative/communicative approach to planning is therefore more of an ethical "worldview" than a paradigm in the sense of Kuhn (1962/1981).

With this in mind, Tewdwr-Jones and Allmendinger (2002) argue—and this constitutes the core of the post-positivistic approach to planning—that we must shift our attention to a foundational, paradigmatic level and, in the spirit of Feyerabend's (1975/1979) critique of Kuhn, realize that "... there are likely to be many competing and overlapping paradigms that are incommensurable in certain aspects and share common understandings in others ... there is not one dominant paradigm in planning today..." (Tewdwr-Jones and Allmendinger 2002, 214f). Hence, Tewdwr-Jones and Allmendinger dismiss the idea that the collaborative/communicative is the only approach to planning alive today. Instead, they advocate a theoretical pluralism in which various paradigms are allowed to flourish and are put into practice by planners.

With this in mind, the argument of Tewdwr-Jones and Allmendinger (2002) immediately leads us to the underlying assumption expressed earlier in chapter two of the present volume: "This manifested itself in the acceptance that knowledge is never wholly reliable and that it always depends crucially on certain metaphysical assumptions or paradigms (see, for example, Kuhn 1962/1981, Toulmin 1972/1978 or Feyerabend 1975/1979). This resulted in planners realizing that concepts such as the "objective" description of a problem, or the "optimal" solution to a problem, could no longer serve as the basis of discussions in planning (nor, indeed, should they so serve), because neither really exists in the first place. Moreover, the significance of values (axiology and ethics) became apparent to planners. ... Hopefully, this choice in nomenclature will serve to clarify that we are not only dealing with the individual people who operate within the "Planning World" (and usually cooperate within some organization), but that we must also take their background knowledge (viz. the methods, concepts, theories, worldviews, and paradigms on which they draw) into account. (For further details, see Chapter two above.)

Since we are here dealing with central assumptions that form the conceptual foundation of this book, it is clear that the post-positivistic arguments are in agreement with the present volume.

To summarize, we may therefore conclude as follows: The three mainstream approaches to planning enumerated above are compatible with the planning model described in this book. They do not contradict one another. Indeed, the same thing could be said for most other approaches to planning, provided they proceed from the following two basic assumptions: a) knowledge is never objective but rather always socially constructed and cannot be had without a knowing subject, and b) there is always more than one point of view according to which we can see the world.

PART II
CONSTRUCTS
FOR THE TREATMENT OF
PLANNING TASKS

Introductory Remarks

In our planning theory of the 'third generation,' the development of a "Comprehension of the Situation" was described as a subtask in planning. In the second part of this book, one particular aspect of this topic will be analyzed in detail. Our core question is: what are the conceptual contents of a planning task and how are these developed? With a view towards this question, the so-called semiotic triangle will be introduced as a "conceptual tool"—one with the help of which the treatment of a planning task's conceptual contents can be both structured and supported.

In so doing, we will discuss some rather subtle details regarding the foundations of the issues at stake here. As a result, not every aspect taken into consideration will be equally relevant to every imaginable planning task.

Some readers may wonder why the so-called communicative aspect of planning, which has dominated planning theory so substantially over the past three or four decades, is not discussed in what follows. The reason is this: in our planning theory of the 'third generation,' a distinction is made between "Comprehension of the Situation" and "Communication about Behavior" (including participation etc.), even though the two cannot be separated. The integration of the communicative aspect into planning is certainly both appropriate and necessary. We must, however, distinguish between the conceptual contents of a planning task on the one hand, and the social, psychological etc. conditions that govern the emergence and treatment of these conceptual contents, to which the topics of communication, discourse etc. belong, on the other hand. These topics are, at least implicitly, sometimes wrongly assumed to be identical. "Communication" and "discourse" influence the treatment of conceptual contents, but are not to be equated with conceptual contents. Fruitful communication in planning is, after all, only possible if the relevant conceptual contents are first developed. The second part of this book therefore deals with conceptual contents, and not with topics like communication, discourse, etc.

Chapter 3

The Semiotic Triangle – A Conceptual Tool in Planning[1]

Introduction

What is a "pedestrian zone"? What is a "region"? What are "slums"? What is "nature"? What is "garbage"? The problem here is as follows: are these respective things material objects, or are they merely thoughts in our human "Cognitive Apparatus?"

If one poses this question, planners—though not only planners—even today often respond as follows: "pedestrian zone", "region", "slum", etc. are obviously material objects. This answer, however, is inaccurate; only a naïve realist (see Bunge 1996, 354) would regard this as incontrovertibly true. "Pedestrian zone", like all the other above-mentioned concepts, are human thoughts, not material objects.

Whichever among the numerous definitions of planning one prefers, at bottom they all concern the idea that something is to be improved about a situation regarded as wanting. This, however, (trivially) presupposes that the planner has an accurate comprehension of the situation in question; fundamental misunderstandings and confusions, such as the one between objects and thoughts, should not be allowed to creep in. Unfortunately, such misunderstandings are by no means the exception. The result: if it is not clear whether the object of investigation is a thought or a material object, the seeds for complication in the further development and implementation of the plan have already been sown.

The following section elucidates the semiotic triangle as a basic scientific construction—one that will both allow us to avoid the aforementioned confusions and aid us in developing a comprehension of situations by acting as a supportive "conceptual tool" in that endeavor. It accomplishes the latter on all levels of planning—from architectural and urban design, to urban and regional planning, all the way to spatial planning at the level of a state.

The topic brought up in this introduction (roughly put, the relationship between language/signs or ideas/thoughts and objects) is, as a philosophical problem, by

---

1   Many scientific and philosophical works focus on analysis. However, treatises which strive for a synthesis are equally helpful to planners if what is presented by means of this synthesis can be made serviceable as a conceptual tool in planning. With this in mind, we will here—at least with regard to certain particular aspects of the theoretical discussion and following Bunge (1974a,b)—attempt such a synthesis. The attendant risk of thereby producing open questions is consciously accepted. To put it figuratively: "First the tree, then the sawdust." (Bunge 1974a, v)

no means new. It was first raised in Plato's dialogue "Euthyphro" (see Ros 1989, 19ff). This topic has, in other words, been discussed for the past 2400 years—which is to say since Plato (428/427–348/347 B.C.), Socrates (died 399 B.C.) and Aristotle (384–322 B.C.)—in philosophy, the theory of science, as well as many other disciplines. Since that time, many notable philosophers have examined this problem from a variety of perspectives, among them Augustine, Locke, Leibniz, Kant, Frege, Wittgenstein, Mead, Carnap, Popper, and Bunge, to name just a few (see e.g. Ros 1989, 1990a, 1990b). The expression "semiotic triangle" is a relatively recent coinage, and is therefore not used by all philosophers and scientists.

Even today this question plays a role in a variety of academic disciplines. In fact—and, given this long history, how could it be otherwise?—many works in the theory of planning and architecture deal with the topic as well (see e.g. Bense 1971).

Be that as it may, an attempt to summarize the essential aspects of this topic as they relate to planning certainly shouldn't be superfluous. This conviction is based on three sources of information: firstly, on the authors own extensive practical experience in planning, secondly, on observation of the education of planners, and finally, on familiarity with much literature on the theory and methodology of planning.

Before delving into a detailed description of the semiotic triangle we shall first attempt to adumbrate, at least in broad strokes, the core of the problem, as well as explain its relevance to planning.

An Illustrative Experiment

One of the fundamental distinctions that is of importance to our investigation can be illustrated by way of a simple experiment. Imagine the following situation (it's best to try this out with acquaintances): you write an Arabic numeral—a "7", for example—onto a blackboard with chalk, and ask: "What do you see here?" Some members of the audience will surely be annoyed by the triviality of the question, some will feel that they're being "taken for a ride", and others may "smell" a trap. In most cases, however, the answer that comes relatively quickly is: "a seven". And obviously most people "see" a seven here.

But is this right? Of course not. Of course one doesn't "see" a seven. Note that the question was not "what is this?", but "what do you *see* here?" What one actually sees is some stuff (chalk) in a particular color and shape on a surface. For example, chalk: white, board: green. (Strictly speaking, one would even have to first come to an agreement about the concept "chalk" or the color "green" of the board.) The shape of the chalk markings caused the audience to "see" a seven. That the figure on the board is a seven is not, however, something one can see; rather, this is something that we were taught as children. We recall what we have learned and create the seven in our "Cognitive Apparatus" at the moment when we perceive such a pattern, and then voice this thought as, for example, an answer to the question "what do you see here?".

The fact that the chalk figure drawn on the board in a particular shape and the natural number 7 are different things becomes even more plain if we write a Roman "VII" next to the Arabic "7", and then perhaps even the word "seven" next to those. All three chalk figures—"7", "VII", and "seven"—look completely *different*, yet most people still say that they "see" the *same* thing in all three. That can't be. By this point it becomes (exceedingly) clear that there is no natural number 7 to be "seen" on the board. It also becomes clear that the natural number 7 has nothing directly to do with the three accumulations of chalk on the board, and therefore isn't at, in, or by the chalk figures on the board either. This means that there is a difference between the signs (the chalk markings drawn on the board in a particular shape) and the things being designated (in this case the natural number 7).

We call this designated—in our example the natural number 7 (which we characterized as a thought above)—a "construct". (Strictly speaking, we are here dealing with a "concept", and concepts are one of the four basic categories of constructs; for details see Chapter 3.1 below.)

If the natural number 7 is not at, in, or by the chalk figures on the board, where is it located? The answer is: constructs, such as the natural number 7, are located exclusively in our human "Cognitive Apparatus"[2]—not outside of it, not on the board, and also not somewhere between the board and our "Cognitive Apparatus". Outside, that is, in the world outside our brains, there are no sevens: there are only seven objects, seven houses, seven trees, seven cars, seven "somethings"—but not seven on its own. What empirically corresponds to the natural number seven is merely a bundle of neurons in our brains.

This thought experiment illustrates some of the essential distinctions that are marked out in the so-called semiotic triangle. The following are to be distinguished:

(a) Language, signs: here, for example, the shape of the Arabic or Roman seven drawn on the board using chalk.
(b) Constructs: the imaginative entities in our "Cognitive Apparatus", such as the natural number seven.
(c) Objects (pieces of chalk, boards, trees, humans, etc.) and events; whereby we are dealing with an "event" if one of the properties of an object changes.

The relationship between language/signs, constructs and objects/events can be represented by a triangle. Figure 3.1 shows the components of the semiotic triangle as well as the relations that hold between them: language designates constructs and denotes objects/events, while constructs refer to objects/events. The following example may serve to illustrate these distinctions: the word "Paris" is disyllabic (← language); the construct "Paris" designates a European metropolis (← construct); in Paris we can find the Eiffel tower, many paintings in museums etc. (← reference to objects).

When we create or use constructs, the following three components are almost always involved: objects/events, language/signs and constructs themselves. Constructs are therefore not isolated entities of thought; they stand in relation to and interaction with our linguistic possibilities, the variety and problems of the objective world, as well as the already existing stock of constructs.

---

2   In what follows, the term "brain" will be used for the brain construed as a biological/chemical/physical entity; by "Cognitive Apparatus" etc. we will mean thought processes.

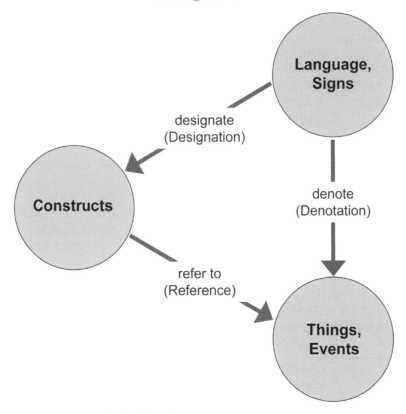

Figure 3.1   The semiotic triangle

Source: After Bunge 1974a, 1974b; modified and simplified

Eco (1977, 30) has compiled a list of the terms used by other authors to designate what we have here called language/sign, construct and object/event. For "construct" the following expressions have been used:

"interpretant" (Peirce); "reference" (Ogden-Richards); "sense" (Frege); "intension" (Carnap); "designatum" (Morris 1946); "connotation" (J.S. Mill); "mental image" (Saussure, Peirce); "content" (Hjelmslev); "state of consciousness" (Buyssens)

Synonyms for "language/sing" are:

"sign" (Peirce); "symbol" (Ogden-Richards); "sign vehicle" (Morris); "expression" (Hjelmslev); "representanem" (Peirce); "sem" (Buyssens)

Synonyms for "object/event" are:

"denotatum" (Morris); "reference" (Frege); "denotation" (Russell); "extension" (Carnap)

Despite the differences in the terms used, all the aforementioned thinkers are really dealing with the same basic structure.

The following text will focus on clarifying the distinctions found in the semiotic triangle. It must, however, be stressed that in working with the semiotic triangle, we are dealing with a constant interplay and flow of reciprocal influences between all three components (language/signs, constructs, and objects/events), although entry is almost always gained by way of language/signs. The three constitute each other. Briefly put: we differentiate the three components, even though they cannot be separated from each other.

Everybody knows, or at least presumes to know, the difference between objects (such as a tree, a house, etc.) and languages (for example "German", "English", etc.) or signs (such as pictograms, the letters of an alphabet, the numerals from 0 to 9, including 7). Only a few, however, have a clear idea about what to make of constructs.

Before we delve further into the peculiarities of constructs, we shall first illustrate, by way of some examples, which constructs are used in planning. In the process, we shall also clarify the point of the distinction between the three components of the semiotic triangle for planning.

What are constructs in planning?

Table 3.1 below lists some of the constructs that are used in planning. The constructs are shown on the left, while to the right of each, several objects are given that, depending on the definition used, may be referred to by the respective construct. Both constructs and objects are of course represented through linguistic signs in the table.

For example, the expression "land use" in the left column *designates* a construct. To put it differently: there is no object called "land use". As a rule, a construct like "land use" refers to certain objects: buildings, earth etc. Such objects are listed in the right-hand column.

Strictly speaking—and anticipating the elucidation offered in Section 3.1—we ought to phrase it slightly differently: the right-hand column contains linguistic expressions that simultaneously *denote* objects and *designate* constructs. Example: the expression "tree" not only denotes an object, it simultaneously designates a construct. What really matters at this point is that the left-hand column contains no objects, but only constructs. The right-hand column, on the other hand, contains objects. That constructs are "hidden" in the right-hand column as well is something to which we shall return in Section 3.1 below.

Take the first expression in Table 3.1. "Garbage" is a construct, not something concrete. The topic of garbage is particularly well suited to demonstrate that it was a redefinition of the construct that produced progress in waste management. In the past, the construct "garbage" referred to objects which were generally either incinerated or buried. Today, the construct has changed as re-use and reduction have been added as new aspects of the construct. Substance cycles are incorporated, ecological assessments established, etc. It was only as a consequence of this change in the construct that new kinds of products were produced, while others ceased to be used because they were, for example, not recyclable.

The way we deal with the question of garbage is an example of how tasks in planning often consist in first contemplating the construct at hand. *At first*, the objects themselves are not changed at all. That is, the objects to which the construct "garbage" refers remain as they are. They are, for example, still the same empty, dented aluminum cans.

Table 3.1   Constructs and objects (see text for explanation)

The linguistic expressions in this table ...

designate constructs	denote objects (and designate constructs)
Garbage	dented can, old tyre
Species diversity	bird, Persian cat, Ebola virus
Land value	five-hundred Euro bill, earth
Preservation of Historic Landmarks	statue, half-timbered house, old building
Town development	marketplace, building
Land use	house, meadow
Pedestrian zone	pedestrian, cobblestone
Industrial park	warehouse, shop floor, duct tubing
Landscape painting	painted canvas, hill, tree
Market	tomato, egg, market stall, monger
Nature	oak, human, AIDS virus
Conservation	moose, Signpost: "Conservation Area"
Region	ground, tree, veal sausage, apple wine
Beltway	tar, manhole
Slum	shanty, plywood dwelling
City center	high-rise, church building, city hall
Urbanism	densely constructed housing, people
Traffic	traffic light, car, person, streetcar
Apartment	stone, windowpane, door handle, tile

An example from a different field: An apartment is also a construct and not an object. This can be seen in that the very same objects to which the construct of an (empty, unfurnished) apartment refers can also be used as an office. Correspondingly, "office" is also a construct.

Constructs like "apartment", "office" or even "street" make it clear that something so fundamental is involved here that we hardly notice it. Quite a bit of intellectual effort is required in order to realize that our perception, our thoughts and our actions are determined by constructs in elementary ways. "In looking at a street, for example, a jumble of variously intersecting, partially moving configurations of stimuli appear on the retina. These are the initial data given to perception. What is immediately seen, however, are houses, trees, shops, cars, people—briefly put, an organized world. We do not thereby feel that any sort of processing is necessary to recognize the order of things in the jumble of visual impressions" (Hoffmann

1986, 12). Rather, the visual stimuli directly activate the constructs correlated with them. "The result of this process is the perception of the given stimuli as conceptual objects. Or, to put it differently: we see our world in the … [constructs] that we have created as behaviorally necessary classifications" (Hoffmann 1986, 12).

Table 3.1 clearly demonstrates that a multitude of the topics that concern planners are in fact constructs. Through this list it becomes clear that in planning we are always, among other things, dealing with two tasks: to plan and to draw up plans is always to work with constructs on the one hand *and* objects on the other hand, both of which happens primarily via language/signs. While plans always aim to improve the objects (among which people are counted) or the circumstances of living objects (people, animals, plants) of this world, this cannot be done without constructs, because it is they that are the bearers of our knowledge (as we shall see below). They offer us insights into situations and open our view to certain relationships; but in so doing, they often also hide from our view other relationships that are not contained in them. In this manner, they dictate what counts as a "fact" and which arguments are accepted as relevant and convincing, and thus ultimately end up determining our actions in planning. The importance of this latter point cannot be overstated.

In planning—as in so many other areas of life—we always use constructs, no matter whether we are aware of it or not, and independently of whether they adequately represent the situation or not. The improvement of constructs is therefore an essential means by which to achieve better results in planning. Constructs are said to be "better" if they (a) do not suffer from internal contradictions, (b) are logically compatible with as many other constructs as possible (which is to say that they manage to integrate as many [partial] constructs as possible), (c) agree with all available factual knowledge, and (d) are furthermore helpful in relation to the planning problem at hand. Constructs that are "better" in this sense will be in use for longer periods of time before they are replaced by newer constructs.

In this section we have, on the basis of a few examples, discussed what sorts of things constructs in planning are. In addition, we have illustrated their significance. These constructs will serve as the main topic of the text follows. In order to be able to integrate this topic properly, it is necessary first to make a few short comments on "language and signs" as well as "objects and events".

3.1   The Components of the Semiotic Triangle

Language and Signs

The importance of language derives from the fact that constructs are expressed and communicated exclusively through language and signs.

Languages are systems of codified signs that are used for the purpose of communication (see Bunge 1974a, 8). Language is the use of signs to comprehend something or to communicate with someone. "Comprehension of the Situation" and communication with those involved in the planning task is the basis of success in planning. "Comprehension is achieved if the representation [of a planning problem] is such that it is clear, accurate, conclusive and elucidatory. Understanding is achieved if the continuation of the exchange of messages [on the level of language] no longer influences the agreement in comprehension [on the level of constructs] of those involved." (Heidemann 1990, 5)

By "language" we usually understand spoken or written language, or human tools of communication in general (see Signer 1994, 49). The topic of language can—albeit in a compromised manner—be explicated as follows (see for example also Eco 1991, Bußmann 1990, Heidemann 1990 or Signer 1994, 53ff). First off, the question of signs.

Berkeley and Pierce, for example, both offered the same definition of "sign": something that stands for something else (see Bußmann 1990, 864). Three types of signs are generally distinguished: icon, symbol and index.

Icon

The relation between the icon and the denoted state of affairs is one of actual, pictorial similarity between sign and material object. Examples are pictograms, like those used to designate different sports or those of a running person used to mark escape routes and emergency exits. Icons are likenesses that either simplify, accentuate, or distort (see Rodi 1992, 297). In the context of planning, it is important to note that building plans (blueprints, plan views, etc.) and cardboard or wire models of buildings (which do not require a legend due to their illustrative character) are to be classed among icons, because they exploit an actual, pictorial similarity between the iconic sign (plan or model) and the material object (building, neighborhood, etc.). Icons can therefore only represent material objects. As a result, iconic representations of, for example, the construct "office" are only able to represent the actual objects, or material parts of the constructs, but never the construct as a whole. This is to say that the means of representation preferred by planners can't represent constructs themselves, but reflect only their material parts.

Symbol

The relation between a symbol and the designated state of affairs is achieved by means of prior agreement, through "semiotic interpretation" (see especially Section 3.8 below). This agreement rests exclusively on convention; the meaning of a symbol is thus generally dependent on culture and language. A sign therefore counts as a symbol if it was correlated with a state of affairs through agreement—if it was, in other words, "baptized" (see Heidemann 1990, 9).

In the context of planning, this means that plans whose relation to the designated object is only possible on the basis of an agreement (through a legend, for example) count as symbols. Most representations in development and land-use plans count as symbols, because the signs appearing in the plan must be "baptized" through agreement.

In other words, when we define the constructs which constitute a planning task, we are undertaking a so-called semiotic interpretation. In the process, we primarily use symbols, and symbols are signs that without exception require semiotic interpretations, since they, by definition, rely on prior agreement. From this we can conclude that even in planning symbols symbolize nothing without semiotic interpretation. A symbol, such as the sign-sequence "city development", must be semiotically interpreted through constructs, otherwise this sequence of signs would symbolize nothing. Without semiotic interpretation "city development" is devoid of content—nothing but an empty sound.

Index

"The relationship between an index and the designated state of affairs does not rely on similarity as with the icon, nor on convention as with the symbol, but is rather apprehended on the basis of a causal connection." (Signer 1994, 79f see also Bußmann 1990, 330) This means that there must be a cause and that it must be recognized as being the cause. The particular course followed by a crack in a wall is, for example, an index for the sinking of the foundation at a particular place, and smoke is an index for fire.

In the following section, symbols will be especially important.

Language

It is not important in which language our knowledge concerning a planning task is encoded; that is to say: it is not important which sign system is used to convey this knowledge. Whether or not a language is useful depends primarily on whether it is able to serve in the process of clarifying constructs.

Regarding the sign systems actually in use, we may distinguish between the following languages (see Bußmann 1990):

- Natural versus artificial languages: "German" or "French" are, for example, natural languages. Programming languages, sign systems for rendering flow charts, or the signs used to represent logical operations are all examples of artificial languages.
- Symbolic versus non-symbolic languages: symbolic languages are languages that use symbols as signs, such as the natural and artificial languages. In non-symbolic languages, like non-verbal communication, for example, no symbols that require "baptism" through prior agreement are used.

Planners use more or less all of these languages.

In the context of planning, the following five kinds of languages can be found: colloquial language, technical language, object language, metalanguage and jargon (for details see Stegmüller 1983, 63ff, Bußmann 1990 of Signer 1994, 58ff). All languages, and especially the transitions from one language to the next, present the danger of slipping into *jargon*, into "incomprehensible murmuring" (Bußmann 1990, 361), in which watered-down expressions are used that lack both clarity and precision.

This brief sketch of language and signs will suffice for our needs.[3] We will now turn our attention to the second component of the semiotic triangle, namely objects and events.

Objects and Events

Putting several special philosophical questions (see e.g. Vollmer 1993), such as whether the "external world" exists only as ideas in our heads, aside for the moment, we may characterize objects thus: objects are the things that make up physical reality. This includes both natural things like trees, stones, or people, and artifacts like houses, paper, plans, or books.

Objects are either specified or placed in relation to other objects with the aid of properties regarding quality, quantity, space, and time. Mass, color, or temperature are examples of properties. Properties can also have differing modulations; the temperature of an object can, for example, be five degrees celsius in one case, and twenty in a different case. The sum of all modulations of every property is called a state. A change in (at least) one modulation of a property is called an event, and a sequence of events is called a process.

Both the existence of an object in a particular state and the occurrence of an event in an object are called a fact. "... a (real) fact is either the being of a thing in a given state, or an event occurring in a thing" (Bunge 1977, 267).

Objects exist in space and time. This means that they have extension and duration. If this were not the case, they could not be perceived through observation. Objects also have the property of being located at a particular place, of having energy, and of being capable of change.

Objects are autonomous entities. They are external to cognitive subjects like thinking human beings (see Bunge 1977, 16f), meaning that they exist outside our "Cognitive Apparatus". If this were not so, they could not be made the objects of analysis, and human thought would be pure introspection. That is, objects will here be understood as autonomous in the sense that they exist independently of the perceiving subject. Accordingly, they are just as they are; they are not dependent upon what we can know or apprehend about them. An example of this are radioactive substances like uranium, which existed long before people were able to measure radioactivity or to isolate and use these elements.

---

3   The topic of linguistic and non-linguistic sign-systems has not been treated exhaustively in this commentary on signs and the various kinds of language (natural, artificial, symbolic, non-symbolic, colloquial, technical etc.). These are, for the moment, just classifications.

In order for language to serve as a means of communication, rules and agreements must underlie its use. Languages are therefore tied to systems of rules that enable a certain group of people to use them. These rule-systems are investigated in linguistics, to which the following subtopics belong:

(a) syntax, which deals with the grammar and repertoire of signs of a language,

(b) semantics, which focuses on meaning in language, and

(c) pragmatics, which deals with the consequences of language in relation to the behavior of those who use it (see e.g. Stegmüller 1983, 68ff, Bußmann 1990, Eco 1991 or, as an introduction, Miller 1981).

Essential to this point is that we humans cannot perceive the objects and events of this world without using constructs. This has been corroborated through numerous analyses in science: "Even the most simple (sensory) perceptions are embedded in cognitive patterns [read: constructs] and inextricably interwoven with them." (Groeben and Westmeyer 1975, 202). Consequently, it became clear through discussions in the philosophy of science that "facts ... are always dependent on implicit observational theories [= constructs]" (Groeben and Westmeyer 1975, 202). That is to say, it is fundamentally impossible to perceive so-called "hard facts" "in themselves", because these facts are always only acquired in a cognitive- and construct-dependent manner (see also Kuhn 1962/1981 and Feyerabend 1975/1979). This is also why the hope of certain "naïve realists" (see Bunge 1996, 354)—the hope of being able to perceive facts without thereby using constructs—has shown itself to be folly. As a matter of principle, constructs are always involved if we try to apprehend or work on a fact. Planning therefore doesn't deal directly with objects or events (i.e. facts), but with constructs about facts—a difference of the utmost importance. Consequently, we cannot avoid an examination of constructs in planning.

Every *factual* change is of course only possible as a change in, or of, objects—where we are, in this case, also counting humans as objects. In solving a problem in planning, it is generally the case that we have to develop a course of action (i.e. a plan) so as to change a constellation of objects that is regarded as somehow disadvantageous. This, however, is, as mentioned, not possible without constructs: the treatment of constructs is an essential precondition for a plan-governed change of objects, because it is constructs that bear our knowledge and determine our actions. In contrast to this, a change in constructs *per se*, without any relation to objects and events, is only a change of thoughts and, in this sense, without any relation to factual reality.

Constructs

What is a construct? A construct is that which is designated by the construct-expression, and is therefore often called its meaning.

In contrast to material objects, constructs are abstract, or fictional, and therefore *conceptual* objects. (For the following treatment see especially Bunge 1974a, 1974b, 1983a, 1996; a concise description of the philosophical foundations can be found in e.g. Bunge 1996, 241ff.)

One of the starting points of our engagement with constructs lies in the fact that we use words in language which, unlike proper names, don't refer to exactly one person or object (for example the "Bismarck-Oak" to the right on the forest path that leads from city X to town Y).[4] Instead, they are words (like e.g. "tree", "birch", or

---

4   Translator's note: after the wars of the mid-nineteenth century, many veterans associations were founded in Germany that erected monuments or planted so-called "Bismarck-Oaks" to commemorate rulers. The point in our context is that the definite description "the Bismarck-Oak on the forest path from city X to town Y" is given as an example of a singular term that, like proper names, refers to a particular thing.

"marketplace") that designate constructs, because they relate to *many* objects (that is to say, all trees, all birches, or all marketplaces).

What makes a construct (like "seven", "tree", "marketplace", "pedestrian zone", or "region") into a construct can be described as follows: if we think of a seven, then we have the construct seven before us once we bracket out

  (a) the language or signs
  (b) the objects or events to which the construct may refer
  (c) the concrete processes in the brain, that is, the processes on the neural level that accompany our image of the seven, and
  (d) the physical process of communication, such as, for example, the changes in the larynx during sound formation, the sound waves, etc.

We say that everything is either a material object *or* a construct, and that nothing is *both* a material object *and* a construct; "… none is both" (Bunge 1974a, 26).

That is, there are certain linguistic expressions—like "birch"—that (firstly) *denote* certain objects (a tree with its branches, the bark, the roots etc.) and (secondly) also *designate* constructs (namely the construct "birch", such as in, e.g., a scientific definition that mentions to which species of plant the birch belongs, the type of its leaves, etc.). This is the case for all the words listed in the right-hand column of Table 3.1 (see p. 62 above).

Then there are linguistic expressions which only designate constructs, like, for example, the number "7" or "urbanism", because there is no object with the name "7" nor one with the name "urbanism." These expressions do not denote objects. The same thing goes for all the words listed in the left-hand column of Table 3.1.

From the perspective of constructs, this same situation can be described as follows: some constructs refer to material objects. That is, some have, or refer to, empirical referents, and others don't. Thus, the construct "birch" refers to all birches, i.e. to objects. As construct "in itself", the natural number 7, on the other hand, does not refer to any objects. In the material world, seven, taken on its own, does not exist; what exists are—as mentioned—seven houses, seven trees, seven "something", but not "seven" on its own. The situation is, for example, similar with the different types of scales frequently used in planning: the nominal-scale, the ordinal-scale, the differential- or integral-scale, and the relational-scale; they too are "in themselves" without empirical referents.

A construct is called "factual" if it is correlated with one or more material objects or events as referents. An example of a factual construct with empirical referents in planning is the construct "degree of motorization". Here, the material objects to which the construct refers are the cars and the people living in a particular area. Another example is the factual construct "individual traffic" which refers to material objects like traffic lights, cars, or people.

Kinds of Constructs

We distinguish four basic classes of constructs: concepts, propositions, contexts and theories (see Bunge 1983a, 444).

(a) "Concepts are the units with which propositions are constructed: they are conceptual atoms." (Bunge 1983a, 44) Concepts are class formations over sets of objects. One obtains the respective set by collecting into a single class all the objects that satisfy a certain property. Concepts are consequently formed by assigning properties to the objects that are to be described. "Apartment", "office", "neighborhood", and "region" are all examples of concepts. All the expressions listed in the left-hand column of Table 3.1 in Chapter 3 above are concepts and therefore belong to this first basic class of constructs.

(b) Propositions (also called "assertions" and "statements") establish a connection between concepts. That is to say, propositions are formed by placing concepts into relation with each other. (On the topic of relations, see Chapter 3.3 and 3.4 below.) For example: in the proposition "numbers are concepts", we find the concepts "numbers" (or "the set of all numbers"), "are" (or "is contained in"[5]), and "concepts" (or "the category of all concepts").

Another example: in the context of planning, the constituent elements, or parts, of the mathematical function describing the outwardly decreasing population density of cities designate concepts. As a whole, however, the function is a proposition, which in turn refers to a stable pattern of objects and events in reality.[6]

Each of the following sentences also express a proposition: "2+2=4", "Pedestrian zones boost the sales of adjoining private businesses", "certain buildings have an intimidating effect on the viewer". There are multitudes of propositions, because almost every kind of assertion or sentence we use day to day expresses a proposition.[7]

A proposition is, in other words, what is designated by a sentence. "Better put: since not every sentence designates, one ought to say that every proposition is designated by one or more sentences." (Bunge 1983a, 58) For example, the sentences "3>2", "III>II", and "three is greater than two" all designate the same proposition.

Note, however, that although every proposition is expressible by one or more sentences, the reverse is not true (Bunge 1983a, 56), since not every sentence designates a proposition. As a matter of fact, there are grammatically well-formed sentences that still don't designate a proposition, like, for example, "the number seven fidgets" or "the square root of a design idea is a song" (analogous to Bunge 1983a, 56). Such sentences are ontologically ill-formed (see Section 3.11 below).

That is to say, propositions are not sentences; propositions are rather expressed by sentences—they are what is meant by the content of the sentence.

In other words: "concepts … are the units of meaning and hence the building blocks of … discourse. We use concepts to form propositions, just as we analyze complex propositions into simpler ones and these, in turn, into concepts". (Bunge 1996, 49) Concepts—like, for example, "car", "neighborhood", or "region"—are therefore the foundational units and thus the building blocks for our work in planning. We use concepts to form propositions. From simple propositions we form more complex propositions,

---

5   This is a so-called qualitative relation; see Section 3.3 below.

6   Mathematical functions are also a kind of relation; see Section 3.3 below.

7   One should note that propositions are not proposals: "note that propositions should not be mistaken for proposals, such as 'let's go'" (Bunge 1996, 49); the same is true for "it was proposed to investigate the logic of problems", etc. (Bunge 1999a, 228)

just as we can decompose more complex propositions into simpler ones, where these latter can, in turn, be decomposed into (and analyzed by) concepts and relations. This, by the way, means that the basic classes of constructs presented here—and this needs to be emphasized—do not constrain the complexity of the entities of thought developed by means of them (on the topic of complexity, see e.g. Rescher 1998).

(c) "A context is a set of propositions that are composed out of concepts with common referents. For example, the set of propositions correlated with [… individual traffic …] is a context." (Bunge 1983a, 44) A proposition without indication of a context does not have a precise meaning. This is to say that only through the explicit indication of the context is it possible to discover all the logical connections of a proposition and thus determine its content.

(d) "A theory[8] is a context *closed* with respect to logical operations. A theory is, in other words, a set of propositions that are logically connected to each other and possess common referents. Example: the theory of evolution by natural selection." (Bunge 1983a, 44).

Figure 3.2 illustrates the four basic classes of constructs.
As the following examples demonstrate, all four kinds of constructs can be found in planning:

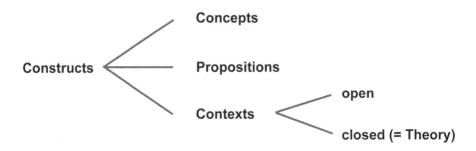

Figure 3.2   Kinds of constructs

---

8   "Few concepts have fared worse … than that of theory. The worst and most popular mistakes in this regard are the following: Theory is any discourse on generalities, however obscure or incoherent; theory is the opposite of hard fact (a vulgar belief); theories are useless: only data and actions are valuable; theories are general orientations or approaches; theories are the same as hypotheses (e.g. Popper); theories are collections of definitions (e.g., conventionalists and Parsons); all theories are generalizations from observed facts (inductivism); there are a priori theories of human behavior (e.g., von Mises 1949); the axioms of a theory are indisputably true (commonsense view); and every axiom system is abstract – that is, uninterpreted (e.g., Debreu 1959, x)." (Bunge 1996, 113) "[A]ll ten concepts of theory listed above are wrong. "(Bunge 1996, 114; consult for details)

(a) All approaches to planning utilize concepts (like, for example, "sustainability", "city", or "city of short distances").

(b) All approaches to planning also include propositions (assertions). The planning approach of "city of short distances", for example, rests on the proposition "… that dense and functionally mixed urban structures with high-quality open spaces and facilities produce less traffic" (Jessen 1996, 3).

(c) Propositions, in turn, have their specific context. Often, it is only a proposition's connection to its context that renders the content of the proposition precise and thus open to scrutiny. Thus, the construct of sustainability is well-suited to the establishment of a metric regarding the use of a forest in the context of forestry. Metrics for "sustainable urban development" cannot, however, be immediately developed by means of this construct.

(d) Beyond this, there are several *theories* in planning, such as the "Central Place Theory", the "Theory of Growth- and Development-Poles", the "Incubator- and Filter-Theory", or the "Theory of Regional Economic Growth" (see Tank 1987, 59ff).

All approaches to planning can, in other words, be related back to the four basic classes of constructs mentioned above.

Because we only have a few theories (in the sense described above) at our disposal in planning, our discussion in this part of the book will focus on the more fundamental constructs: concepts and propositions (including their context).

In this section we described the basic classes of constructs: concepts are joined—via relations—to form propositions. A number of propositions composed of concepts with common referents produce "open" or "closed" contexts. Closed contexts are called theories. The following chapter will describe how concepts, the first class of constructs, are formed.

3.2   The Formation of Concepts

The determination of the content of concepts (including, as the case may be, the description of the corresponding objects/events, which is to say the specification and determination of the objects of interest) is central to every planning task. The choice and definition of concepts does, after all, influence our actions in planning (see especially Section 3.9 below).

One of the most important procedures in the formation of concepts is assigning properties to an object.[9] However, since we are dealing with two kinds of "objects" by this point—namely constructs (concepts, propositions etc.) and material objects—it is necessary to distinguish two kinds of properties. On the one hand, there are the properties correlated with constructs, which we will call attributes.[10] On the other

---

9   See also e.g. DIN 2330 regarding this.

10   Regarding the question of what types of attributes there are and the relationships in which they stand to one another, see, by way of an introduction, Eysenck und Keane 1998, 233ff.

hand, there are the properties correlated with material objects, for which we will maintain the term "properties". This leads to the following distinction:

A person S possesses property F.

The attribute A represents the property F of S.

The first sentence concerns the material object itself and the second concerns its representation at the level of constructs (see Bunge 1977, 58f). In the case of the first sentence, we can argue about whether person S really possesses property F. In the case of the second sentence, we can argue about whether the chosen attribute A corresponds to the property F (see Signer 1994, 97).

What is essential in this regard is that—as noted in Section 3.1 above—we cannot apprehend the properties of material objects directly, "in themselves", but only by means of the attributes of constructs, that is, the concepts that we develop at the level of constructs.

Predication: Assigning Attributes

Given this distinction, we are now in a position to assign attributes to objects.

All objects have attributes. The basic principle of concept formation is simple: we generalize on the basis of particular cases and examples by retaining their commonalities and excluding their differences, declaring the latter irrelevant in the process (see Eysenck and Keane 1998, 235). The basis of concept formation therefore rests on the attributes that are common to, or "invariant" across, a set of objects. We regard these as invariance-attributes (variable only at boundaries). By establishing this invariance, the object-attributes (cleansed of coincidence) become defined and thus accessible to thought; they are the characteristics of objects and situations relevant to knowledge (see Klix 1992). A child may perceive many different trees, or the same tree many times over. What is invariant across all cases forms the concept "tree". Thus, the concept "tree" is essentially determined by attributes like trunk, bark, branches, twigs, and leaves or needles, and the concept of "pedestrian zone" by its characteristic attributes, such as: "can only be accessed on foot", "exceptions (motor vehicles): ambulances, fire and police vehicles, sanitation vehicles, delivery vehicles at designated times, etc.", "further exceptions: skateboarders, inline skaters, people in wheelchairs", etc. Similar things go for the invariance-attributes of events and processes like driving a car, building streets, drawing plans etc. This assignment of attributes to objects is called "predication" (see Bußmann 1990, 597). Objects are described with regard to quality, quantity, space, time, etc., or placed in relation to other objects by means of predication. Predication is therefore also the foundation and starting point for any form of proposition.

Feldtkeller (1989, 85) gives an example for the process of rendering the concept "wall" precise. "Wall" is, first off, a linguistic entity consisting of four letters. The concept "wall" refers to material objects like wood, stone, glass, metal etc. but is itself not material. Wall is a linguistic expression for something that possesses the following attributes, in which parts of our knowledge about walls are collected together:

- insulation, gradual;
- heat retention, gradual (transformation of light radiation into heat);
- reflection of heat radiation from within;
- permeable to diffuse daylight, controlled;
- permeable to direct sunlight, controlled;
- permeable to UV radiation;
- permeable to cosmic microwave radiation;
- light reflection and absorption, gradual;
- visual connection, controlled;
- possibility of passing through, controlled;
- protection against vermin;
- ventilation, admission of summer breeze;
- blockage of foul air;
- protection against wind;
- blockage of precipitation;
- regulation of humidity, steam diffusion;
- protection of the wall itself against water damage through precipitation or other sources;
- sound dampening, gradual;
- sound reflection and absorption, gradual;
- resistance against mechanical stress;
- fire protection." (Feldtkeller 1989, 85)

If one adds a further attribute to the concept thus described (i.e. to the list of attributes), for example:

- permeable to radon, gradual;

then one has changed and expanded the concept "wall". Assertions about real walls may be made by means of this concept "wall".

Furthermore, attribute combinations may be formed from given attributes by performing logical operations. For example:

- through conjunction ("a and b", someone is employed and female);
- through disjunction (inclusive or) ("a or b", either one, or the other, or both; example: someone comes to work by car, or by train, or both: Park & Ride);
- through exclusive disjunction (exclusive or) ("a or b", either one or the other, but not both; example: someone is either male or female, but not both).

Negations ("not a"; someone is not employed), on the other hand, do not define any attributes (see Bunge 1996, 18). An example: the sentence "this exhibit is not an exhibit of the old kind" contains only a negation and therefore defines nothing.

Consequences for Planning

Concepts, as one of the four basic classes of constructs, have various attributes, of which three are particularly important for our purposes.

First: one attribute of concepts is that they cannot be "true" or "false" (on the definition of the concept "truth" see, e.g., Bunge 1974b, 81ff or Groeben and Westmeyer 1975, 142ff).

"... only propositions can be tested for truth. Concepts cannot be so tested, because they neither assert or deny anything. Hence there are no true or false concepts: concepts can only be exact or fuzzy, applicable or inapplicable, fruitful or barren." (Bunge 1996, 49) Definitions of concepts are therefore merely agreements or conventions (see Bunge 1996, 69). This means that one can, for example, invent whatever concept one pleases. As long as one doesn't enter into contradiction, no one can refute it as "false"; it can, at best, be found uninteresting and thus ignored. Since the criterion "true/false" doesn't apply to concepts, discussions concerning "true" or "false" concepts usually end with the observation: "it depends what is meant by X (= the concept in question)". Whether or not a definition of a concept is regarded as "true" instead depends on whether or not the majority of those involved and/or those in power regard a definition of a concept as "true".[11]

This arbitrariness of concepts is often exploited in planning. Because there are no "true" and "false" concepts, there is always the possibility of just redefining concepts, thereby "defining away" the planning problem at hand, letting it "disappear" so to speak—and this without anyone being in a position to say that the newly chosen definition of the concept is "false" *per se*. The following examples show how the procedure of defining concepts can be used to steer the planning task in some "favored" direction:

Example: housing shortage

Whether or not local politicians take action against a lack of housing depends, in part, on the number of people in the city who seek housing. If their number is great, this creates political pressure to do something that will alleviate the housing shortage—for example, to zone for new residential areas or to financially promote the building of new homes. The figure officially given for the number of people seeking housing was for a long time suspiciously low in many German cities. The explanation: in some cities, only those who were officially registered in the index of the housing authorities were counted as "persons seeking housing". And who was registered as a "person seeking housing" in the index of the housing authorities? Only those who already had housing in the city. That is, the concept "person seeking housing" was defined by two attributes: "is registered in the index of the housing authorities" and "already has housing in the city." Such a definition of the concept leads to an underestimation of the actual number of people seeking housing in cities to which many people would like to relocate from outside.

---

11 "Definitions [of concepts] are stipulations, or conventions, not assumptions. They are true by convention, not by proof or by virtue of empirical evidence." (Bunge 1996, 69)

Example: Drinking Water Pollution

A few years ago, the EU set very low limits for acceptable pollution levels of drinking water—virtually zero-values. Only 0.1 micrograms of any single pesticide and only 0.5 micrograms in total of several pesticides combined were allowed in one liter of water. Thousands of tons of pesticide and fertilizer are, however, still put on fields, meadows and vineyards in EU countries. Accordingly, more than half of all areas used for agriculture in the EU have groundwater polluted by pesticides. The situation is to be redressed by revising the guidelines and increasing the maximum allowed values. That is, the concept "drinking water" will be redefined. The residue won't be reduced, but will be made legal.

The fact that the redefinition of concepts is also used to solve problems outside spatial planning is shown by the following example:

Example: Pension Reform

> The head of the department of labor is preparing a reform that resembles a magic trick. He will keep his promise to once again index pensions to net pay. At the same time, he will still manage to drastically reduce the actual pension payments. If [the head of the labor department] succeeds, pensioners will have to pay billions as victims of the reform. This plan should make retirement insurance shipshape for the next generation. The trick goes like this: the [head of the department of labor] … just wants to redefine the term "net pay" in the formula. All private retirement savings in the form of policies, from life insurance to pension funds, will now likewise be deducted from gross pay. The more these policy holders do (thanks to state support) to secure their retirement, the lower their "new" net pay will be. (Sauga 2000, 28)

Such approaches to "problem solving" are possible because concepts cannot, as explained above, be "true" or "false". Concept definitions are, after all, conventions. One can complain about the *consequences* of these new definitions, but the new definitions cannot be "false" in the sense of "untrue" (unless they are defined in a contradictory way).

A second attribute of constructs is the following: concepts are, as described, formed through an assignment of attributes. The number of attributes that can be used to specify concepts is, in principle, limitless. Consequently, there is always the possibility of characterizing a concept further through additional attributes. Any attempt to define a concept exhaustively and comprehensively is thus doomed to failure. For this reason, it is impossible to define the concept "city" once and for all—there are, after all, infinitely many attributes that can play a role here.

Since the number of possible attributes is, in principle, limitless, one has to decide with which attributes a concept should be defined; and for this choice, an appropriate criterion is needed. A goal-oriented choice of attributes can be pursued if it is clear to what use the concept will be put: in planning tasks, concepts can be defined in relation to the question at hand—attributes are then chosen so as to be appropriate to the solution of that particular question. If such a criterion is lacking, there is a danger that either too few or too many irrelevant attributes are developed and investigated. That is, the definition of a concept used in planning can, as a rule, only be suitably achieved in relation to the question at hand. This, of course, presupposes that the question at hand is formulated in a sufficiently clear manner.

In this context, one of the problems we face in planning (see also Dörner 1976) is that the degree of resolution of the attribute- or property-characteristics can be set to different levels: for a child, the definition of "tree" given earlier may suffice. Landscape planners, on the other hand, will collect subclasses of trees together into sub-concepts—species of birch, for example, will be further subdivided according to the properties of the bark, the shape and length of the leaves etc. Botanists may differentiate the same concept according to geological or geographical criteria, or may recognize the presence of disease or phytopathological degenerations in growth and classify according to these (see Klix 1992). One of the tasks in planning is to choose the degree of resolution that is most appropriate to the question at hand. If, for example, one wants to address the organization of local traffic, one can try to do so (without much chance of success) at the level of the physical or chemical properties of the materials used in the construction of the vehicles (metals, plastics etc.). One will, however, presumably fail because the question has been approached from an overly fine-grained perspective.

Finally, this means that the formation of attributes or properties can be taken as far as one pleases—in the limiting case, all the way to the properties of electrons, neutrons, quarks or even smaller particles (see Bührke 2000, 62f). It thus becomes plain just how "open" the development of concepts really is: there are no "objectively" adequate descriptions of objects, only ones that are adequate to the problem, which is to say suitable in the context of the question at hand.

A third point: attributes of course don't just play a role in the opening phases of planning, that is, in the definition of concepts. They also appear at subsequent stages of work, albeit often under different names. For example, that which is called "attribute" during the phase in which we define concepts, is called an assessment criterion during the application of an assessment procedure. Criteria are therefore a subset of all the attributes of an object. More precisely put: criteria are those attributes which are chosen from among the set of all attributes for the purpose of comparing objects (see Strassert 1995, 30). That is to say, all criteria are attributes, but not all attributes are criteria.

In the same way, the attributes defined during the formation of concepts occur again in the determination of goals, specifically as attributes of the state that is to be achieved. Similarly, one encounters them in heuristic creativity techniques (e.g. brainstorming, mind mapping) as so-called parameters, for which different modulations are sought (for example, in the so-called morphological matrix, see Zwicky 1966), while in the evaluation of planning results they are called something else still, namely evaluative criteria.

Judith Innes clarified the central importance of concept definitions in the treatment of planning tasks in her dissertation at MIT. She investigated various cases of planning in order to understand how and under what conditions information influences decisions in planning through "indicators"[12] (and note that almost all

---

12 Indicators are observed states of affairs that in specific instances allow us to posit unobservable states of affairs.

indicators are concepts). As a whole, planning decisions were of course determined by a variety of factors. In her opinion, however, the most important point is that an indicator "… that is socially constructed in the community where it is used" (Innes 1995, 185) influences the respective planning decision to the greatest degree: "to the extent that the users of the indicators had negotiated and agreed on their definitions, they paid attention to what the indicators showed once they were applied." And further: "Knowledge was linked directly to action without the intervening step of decision. Action often simply occurred once there was an agreement on the indicator and a shared understanding of the problem it reflected. Learning, deciding, and acting could not be distinguished. The … stepwise process, assumed by the model of instrumental rationality, where policymakers set goals and ask questions, and experts and planners answer them, simply did not apply…" (Innes 1995, 185).

Innes' discussion reveals how much in planning is already decided with the definitions of concepts: in planning, it is often not so much the formulation of goals or the conscious assessment of a situation that determines the content (and thus the core) of the work. Rather, what is often decisive is the fact that indicators (read: concepts) "are socially constructed in the community" (Innes, see above). Once one has agreed on specific definitions of indicators, one thenceforth only takes note of those indicators and acts accordingly. A separate step of assessment and decision (i.e. the step to which most planning theories ascribe the greatest importance) does not take place any longer, because there is no longer anything to assess or decide. The relevant decisions have already been made (though often without this having been noticed), namely during the step of the process in which indicators, that is concepts, are defined.

In this section we adumbrated how concepts are formed through predication. Furthermore, several consequences salient to planning were pointed out. What we have not yet considered is the topic of relations. Propositions—as described—relate concepts. They are formed by having concepts joined into assertions via relations. The following section lists a series of relations.

3.3 Relations

Concepts are joined together into propositions with the aid of relations.

In principle, there are a whole series of relations that play a role in the formation of propositions, such as the following (see Bunge 1983b, 305ff):

(a) Spatial Relation
Example: city A and city B lie forty kilometers apart.
(b) Temporal Relation
Example: city A was founded one hundred and twenty years before city B.
It takes two hours by train to reach city B from city A.
(c) Qualitative Relation
Example: all A's are B's. All houses are structures.
(d) Statistical Relation
Example: assertions of the kind "X percent of A's are B's". There is a

strong correlation between soil sealing and frequent flooding.
(e) Functional Relation
     Example: assertions of the kind "a thing A reaches point B at time t".
(f) Probabilistic Relation
     Example: the probability that A reaches place B by x o'clock is y.
(g) Relation of Descent (biological principle)
     Example: species A and B have a common ancestor C. Children
     have parents.
(h) Causal Relation (information regarding the causes of events)
     Example: dissatisfaction leads to rebellion. PCB and lindane weaken
     the immune system. Air pollution leads to allergies.

The first two relations convey information regarding space and time. The third, fourth and fifth—(c) to (e)—make assertions about "what goes with what" (Bunge 1983b, 305). The sixth (f) is a special case of the fifth (i.e. the functional) relation. The last two—relations of descent and causal relations—provide information about "what comes from what" (Bunge 1983b, 305; see also Bunge 1996, 18ff).

In the present context, it would take us into too great detail to discuss the particulars of each respective relation (for details see e.g. Bunge 1983b, 305).

In the following discussion, however, the concept of a "mechanism" will be of importance. It is defined as follows: "mechanisms" are a combination of probabilistic (f, above) and causal (h) relations. The following section—3.4—is devoted to this topic.

3.4 Mechanisms

Introduction

Planning generally concerns the attempt to improve a situation that is deemed to be disadvantageous. While working on any planning task, we must always take numerous particular aspects of the task into consideration, many of which were more or less exhaustively discussed in the previous sections.

This Section focuses on a further, fundamental step that must be accomplished when approaching any planning task. Once a disadvantageous situation has been identified and it has been decided in what respect this situation should be changed, the following question presents itself: by what means can this change be achieved? In other words: what will lead to this change? What will bring it about? —however you want to express it. More precisely put: which "mechanisms" can be deployed so that the disadvantageous situation will improve to become the favorable one we desire?

We will call that which brings about the change of one situation into another a "mechanism" (note that this concept has nothing to do with "mechanical"[13]).

_____

13 Our use of the term "mechanism" derives from Bunge 1999b (see also Hedström and Swedberg 1998). (See below for the definition of the concept "mechanism"). This concept ought not to be confused with "mechanics" or "mechanical" (see d'Aboro 1939).

Abstractly put (and anticipating the results of the discussion that follows), we may say that mechanisms can be either causal or probabilistic in their structure, or a combination of both. With regard to causal mechanisms, two basic types can be distinguished: those of Type 1 involve a transfer of energy, while those of Type 2 involve a triggering signal, as in the case of an oral or written order. We shall furthermore distinguish between essential and non-essential mechanisms. Essential mechanisms are what constitute a system at its "core"—we call these simply mechanisms. Most essential mechanisms are affected by non-essential mechanisms, which we call "forces." Forces alter the speed or manner of operation of an essential mechanism in a system. With respect to the semiotic triangle, we will furthermore distinguish between mechanisms and forces on the one hand, and "explanations" on the other hand: the concepts mechanism and force denote what factually exists; explanation, on the other hand, designates the description of the respective mechanisms or forces at the level of constructs. Since (as is the case with concepts) what factually exists can't be directly perceived "in itself", we effectively always work with explanations which we presume (or at least hope) to describe the respective mechanism accurately. Because mechanisms can be treated according to different degrees of resolution (with almost arbitrarily fine grain, for example), it is necessary to choose the degree of resolution in such a way as is appropriate to the question at hand.

The following discussion of mechanisms essentially follows the exposition of Bunge (1999b).

What Role do Mechanisms Play in Planning?

Mechanisms play a decisive role in planning. Without them, there would be nothing that could bring about change, ergo it would be impossible to intervene in systems on the basis of planning. Consequently, the entire profession of planning would be ineffective and thus superfluous by definition. The assumption that mechanisms exist is therefore a *conditio sine qua non* of all planning (although it is not an assumption shared by all theorists[14]). For us the question therefore isn't *whether* mechanisms underlie planning, but only *which* do. This goes for any kind of planning.

---

14 Some theorists dispute the existence of mechanisms. Since most mechanisms are concealed (see below in this Section 3.4) their existence is a matter of conjecture. "Consequently, no self-respecting empiricist (or positivist) can condone the very idea of a mechanism. In fact, consistent positivists in the Ptolemy-Hume-Comte-Mach-Kirchoff-Pearson-Duhem-Ostwald-Watson-Bridgman-Skinner tradition are descriptivists: they reject explanations in terms of hidden mechanisms, in particular causes, regarding them as metaphysical misfits" (Bunge 1999b, 28f). Instead of (also) concerning themselves with mechanisms, empiricists (that is, descriptivists) prefer to concern themselves with the description of observable facts and the connections between directly observable variables (like e.g. inputs and outputs), and distrust generalizations that go beyond the given data. Especially since David Hume, they have disputed the existence of causal mechanisms and have instead defined these merely as a constant conjunction or sequence. The fact that this empiricist point of view is a rather restricted one is shown by the following example: imagine what would have happened if Newton had avoided postulating unobservables like mass or gravitation and their effects on

In what follows, we give several illustrative examples of mechanisms as they are currently discussed in the literature of urban planning. (There are, of course, more complex theoretical approaches (see e.g. Wegener 1994, 17ff; Bossel 1994; Krätke 1995; von Böventer and Hampe 1988; Tank 1987 or Winkelmann 1998, 51ff), though these tend to play a rather more subordinate role in practical urban planning[15]). The respective mechanisms will not be described in their detailed structure (A causes B, B causes C, C causes D etc.) but in abbreviated form:

"New York is currently experimenting with large signs in the streets that measure air pollution and that (as has already been successfully tested in Sao Paulo) move commuters to switch to trains or buses." (Mönninger 1999, 19)
Mechanism: displaying air pollution values causes commuters to switch to trains and buses.

"The demonstration of exemplary construction projects and model projects in the context of a Construction- and Planning-Exposition Northeastern Berlin, in which urban construction, landscaping, planning- and conversion-processes are taken as integral components, could become the driving force behind forward-looking developments, thereby lending both exposure and popularity to Northeastern Berlin as a place of innovation. An increase in appeal through public traffic- and landscaping-projects and a liberal permit policy can be used to motivate private initiatives and investments."(Kunst 1998, 210)
Mechanism: Exemplary construction projects, model projects in urban construction, landscaping, public traffic- and landscaping-projects and a liberal permit policy cause private investors to invest in Northeastern Berlin.

"… the plan assumes … that the urban function of the former Leipziger Street [in Kassel] will become so attractive as a result reestablishing the historical city squares, that it will attract streams of purchasing power which go beyond the endogenous demand of the area" (Hellweg 1998, 283).
Mechanism: attractively designed historical city squares will cause consumers that don't live in the area to shop in local stores.

In order to recommend causally efficacious courses of action, one must have at least a rudimentary understanding of the system (be it an organization, a traffic system, a neighborhood etc.) in which one wishes to intervene. One must have as accurate as possible an idea of how the system "functions". It does not suffice to apprehend and describe the changes taking place within the system as mere sequences of events. Further information is required about *what* it was that *brought these changes about or prevented certain desired changes from occurring*. And how are such changes explained? By laying bare the mechanisms that "drive" the system in question. As long as the mechanisms are unknown, no effective interventions are possible— save perhaps by coincidence or luck—because one is essentially acting "blindly"

---

each other, and had instead confined himself to observable properties and events and their statistic correlation.

15 This assertion is based on the author's own extensive experience in practical planning.

and without knowledge of the relevant causal connections. This is the case for all interventions by means of planning, be they in organizations, traffic systems or neighborhoods. As stringent as possible a description of the mechanism being used is therefore part of any professional planning operation.

Our concern with mechanisms consciously implies a departure from "methods", "instruments", "measures", "mission statements",[16] "strategic models", etc., for which it is not even rudimentarily clear what they effect and how they achieve these effects in particular cases (see e.g. ARL 2000, 9). Of course, no such understanding will ever be completely exhaustive. We are, after all, generally dealing with an agglomeration of mechanisms that are connected via causal chains or causal networks with feedback loops, and therefore ultimately achieve a high degree of complexity. Accordingly, planners usually open themselves up to attack if they propose using particular mechanisms. All the same, if planners want to bring something about in practice, they must at least have some well-founded assumptions about the mechanisms in question. Therefore, rather than proposing some sort of "methods", "instruments", "measures", or (as we shall call them in Section 3.12 below) "rules in planning" that rest merely on short-lived fads or uncritical tradition, it should always be disclosed what mechanisms are used in planning. This demand is undoubtedly difficult to meet. The alternative, however, is even less desirable: if the mechanisms are unknown, then it is, by definition, left completely open what a certain action has brought about in terms of benefits or harm.

In this context, one of the central properties of mechanisms is that they are concealed. They must therefore be expressed (and thus disclosed) through spoken or written language or signs if they are to become intelligible to third parties. Only this kind of a disclosure makes it possible *to discuss* whether what we are attempting to achieve can indeed be achieved with the mechanisms in question.

Given that, in the end, it was always mechanisms that have brought about the disadvantageous situation under consideration, these mechanisms generally also constitute the starting point for any planned intervention. Proposed planning measures therefore often consist of an attempt to either neutralize the relevant mechanisms or to use other mechanisms in order to achieve some desired state.

Accordingly, there is a correspondence in planning between the conjectured mechanisms on the one hand and the proposed courses of action on the other hand: what courses of action are recommended for a specific planning problem depends on which mechanisms are presumed or alleged to be its causes. The following examples ought to clarify this:

- The mechanism of air pollution can explain the rise in respiratory disease.

---

16 "Mission statements remain largely ineffective so long as they remain on a general and imprecise level. For this reason, catch phrases like 'humane city', 'city of short distances', or 'city of urban rationality' are also largely ineffective. Although they provide a sense of where the journey is headed due to the considerable leeway they allow to interpretation, the weighing of competing goals that is always connected to good planning, *along with a presentation of the possible outcomes and the measures needed to realize them, are usually absent.*" (Eppinger 1998, 224; italics not in original)

- The mechanism of subsidization of agriculture in Europe can explain overproduction in certain sectors of European agriculture.

- The mechanism of functional division between places for working, living and recreation can explain the increase in traffic between these places.

The courses of action that correspond to the above mechanisms are listed below, and serve to demonstrate the close ties between mechanisms and courses of action:

- If something is to be done against the spread of respiratory disease, and if increasing air pollution is seen as the triggering mechanism, then the measures that are considered will revolve around the reduction of air pollution.

- If the overproduction of European agriculture is to be reduced, and subsidies are seen as the triggering mechanism, then planning measures will focus on a reduction in subsidies or a change in the mode of subsidization.

- If something is to be done against the increase in traffic, and the functional division between places for working, living and recreation is seen as the triggering mechanism, then something will be undertaken against this division.

What are Mechanisms?

Let us begin with an explanation of the structure of mechanisms. A mechanism is something that either *brings about* or *inhibits* certain changes in the states of concrete systems[17] (neighborhoods, organizations, etc.).[18] All concrete systems are usually equipped with several mechanisms that either promote or inhibit (or block) changes of the system. Mechanisms are thus what enable a system to "function" (or what inhibit the "functioning" of a system).

The difference between a mechanism and a process is the following: processes are descriptions of changes as sequences of states of concrete systems. Mechanisms, on the other hand, incorporate these descriptions of changes, but furthermore also explain what it was that brought these changes about.

There are many categories of mechanisms: electromagnetic, nuclear, chemical, cellular, intercellular, ecological, economic, political, social etc. Of the mechanisms relevant to planning, most are of the social, ecological, economic or political kind, whereas almost none are electromagnetic, nuclear, chemical etc. Among social

---

17 A concrete system is a collection of real things that (a) is held together by certain bonds, (b) behaves as a unit in certain respects, and (c) is embedded in an environment (excepting the universe as a whole). Atoms, molecules, crystals, cells, multi-cellular organisms, eco-systems, and cohesive social groups (like families, companies and whole communities) are all *concrete* systems. Material artifacts (transport systems etc.) are likewise *concrete* systems. On the other hand, theories, classifications, etc., are *conceptual* systems; and systems of signs, like languages for instance, are *semiotic* systems. In contrast, mere collections of objects, even if they are all of the same kind, are not systems because they do not cohere. For example, cohorts, groups of like income, or social classes are not social systems but aggregates. (For details see e.g. Bunge 1979 or 1999b, 17ff.)

18 What we are designating as a "mechanism" here, Dörner calls an "operation" ."An operation is anything that can change a constellation." (Dörner 1995, 297)

mechanisms we can count e.g. the exclusion (or inclusion) of certain people from a group, cooperation and conflict, participation and dissociation, or imitation and innovation.

Kinds of Mechanisms

Mechanisms can be causal,[19] probabilistic,[20,21] or both. Accordingly, explanations may be formulated using the concepts of causation, randomness, or a combination of both.

---

19 The conception of causality that underlies this section is described by Bunge 1987. He criticizes a series of doctrines, among them the total rejection of causation and its equation with regular succession or constant conjunction, as well as the conception according to which causation is the only form of determination. In their stead, he primarily proposes the following theses:

"1. The causal relation is a relation between event—not one between properties or states, let alone one between ideas. More precisely put, causation is not a relation between things (i.e. there is no *causa materialis*). When we say that an object A causes an object B to do C, then we mean that a certain event (or set of events) in A produces a certain change C in the state of B.

2. In contrast to other relations between events, the causal relation is not an external relation between them like the relation of temporal succession: *every effect is somehow brought about by its causes*. Put differently, causation is a kind of event production, or, if one prefers, a kind of energy transfer.

3. *The causal production of events is lawful* and not unpredictable, i.e., there are causal laws, or at least laws of causal status (it is because of this that regularities must be distinguished from statements of law as represented by differential equations).

4. *Causes can modify statistical tendencies* (particularly probabilities) *but are not tendencies themselves*. In the expression "event X causes Y with probability P" (or: "the probability that X causes event Y equals P") the terms "cause" and "probability" are not interdefinable. Furthermore, strict causality is not stochastic.

5. The world is not strictly causal, even though it is determined: not all events connected amongst each other stand in causal relation to each other and not all regularities are causal in nature. Causation is therefore only a variety of determination. Determinism therefore doesn't need to be narrowly construed as causal determinism. Science is deterministic in a weak sense, as it merely demands lawfulness (of any kind) and the absence of miracles." (Bunge 1987, 401f).

20 "Even coincidence, which is at first glance the exact opposite of determination, has its laws, and coincidental events arise from already existing circumstances ... Thus, the appearance of "heads" in a coin toss is by no means a "lawless" event or a "becoming out of nothingness", but the determined result of a determined operation. It is just not the only possible result, or, differently put, the result is not "well defined."" (Bunge 1987, 14)

21 "Randomness is measured by probability." (Bunge 1999a, 237) "Randomness [synonym: chance]. The particular kind of disorder characterized by local irregularity (e.g. individual coin tossing) combined with global regularity (e.g., long-run equal chances of heads and tails.)" (Bunge 1999a, 237) "Chance event = event belonging to a random sequence, i.e. one every member of which has a definite probability. Examples: ... random shuffling of a pack of cards, random choice of a number ..." (Bunge 1999a, 37)

"[C]hance is objective: random events have definite propensities independent of the knowing subject. These objective propensities have nothing to do with uncertainty, which is a state of mind." (Bunge 1999a, 37)

For example, if the migration of people away from rural areas is explained by a lack of work, then we are dealing with a *causal* explanation. If, on the other hand, the heterogeneous composition of a given collection of people or things is explained as a random meeting or as the result of a random sample, then this is a *probabilistic* explanation. Whereas the evolutionary explanation of the specialization of organisms, using concepts like random mutation, crossover, symbiosis, geographic isolation and other factors, is *both*, i.e. hybrid.

The proposed definition—namely, that mechanisms can be causal, probabilistic, or both—avoids the dichotomy of "causal" (or "deterministic") versus "probabilistic", because there are many mechanisms in which both occur in combination.

A simple example may serve to illustrate this point: Schimank (citing Boudon 1984) stresses "... that social processes are strongly influenced by 'Cournot-effects': that is, by independent causal factors whose concurrence has a coincidental character ... as in the event of the wind loosening a shingle at exactly the moment when a pedestrian walks along the sidewalk in front of a house, leaving the shingle to land on his head. The fact that the in themselves complex causal sequences 'wind loosens shingle' and 'man walks along sidewalk' should become interlocked in such a way that the event 'shingle falls on man's head' results, is a contingent event. It is neither impossible nor necessary. Sometimes, such 'Cournot-effects' are highly improbable. In other cases, they can be very probable. The latter is the case if one of the causal factors is, so to speak, latently 'laying in wait' for an extended period, and it is only a matter of time until it coincides with the other causal factor. Perhaps many shingles are loose on a certain roof, and many pedestrians pass by on the sidewalk, and windy days occur frequently. Social processes are rife with such 'Cournot-effects' ..." (Schimank 1996, 24) In the case of mechanisms, we are therefore often dealing with constellations of events that are brought about through the concurrence of initially independent complexes of causes.

Basic Types of Causal Mechanisms

Two types of causal mechanisms ought to be distinguished:
   - Type 1 involves a *transfer of energy*, as in manual work or a collision.
   - Type 2 involves a *triggering signal*, as in an oral or written order to do something.
In the first case, what is decisive is the amount of energy that is transferred. In the second case, a negligible transfer of energy can initiate a process that ultimately requires or consumes a lot of energy. Both types of causal mechanism can therefore be described as weak and strong energy transfers. It must also be noted that there is no such thing as information without a physical bearer, and therefore no exchange of information whose energy consumption equals zero. In the case of the second type of causal mechanism, the effect can be disproportionate in relation to the trigger. That is to say, a relatively small trigger can initiate a process that ends up producing a catastrophic effect—something that is especially easily achieved in unstable systems, as with the proverbial scream that can set off an avalanche. Causal mechanisms of the second type are especially prevalent in social systems, because all social systems are equipped with systems of communication.

A causal mechanism is activated by events which we call causes. These causes may be either external or internal; that is, they may be either environmental stimuli or internal events. Environmental causes can be either physical, social, or a combination of both—such as when a sound wave transmits a work order, that, when heard, effects a thought process, which in turn initiates and directs certain actions.

The Degree of Resolution and Transparency of Mechanisms

The problems that emerge from this definition of mechanisms are obvious: complex situations like the ones dealt with in planning are generally composed of numerous mechanisms, energy transfers, and triggering signals. This means that we almost always simplify things and call something a "mechanism" that, upon closer inspection, is actually a process which is itself composed of numerous particular mechanisms.

In planning, it is therefore necessary to choose a degree of resolution for describing mechanisms that is appropriate to the question at hand—i.e. one that is not too "finely" or too "coarsely" grained.

An example: if infrastructure planners check whether there are enough schools in a given region to guarantee a sufficiently high level of education in the population, then it may suffice to say that "leaning" is the primary mechanism of education in a school.

If, on the other hand, a planner is commissioned to build a school, then he will distinguish various forms of learning (each with its own mechanisms), such as, e.g., "lecturing by the instructor", "project work", "group work", "individual work", etc. He does this because the different forms of learning will often place different demands upon the spaces necessary for each.

Alternatively, a psychologist who wants to investigate the various degrees of retention achievable through each form of learning will want to differentiate the mechanism "learning" in a different and more detailed manner—he might, for example, differentiate it into "classical conditioning", "instrumental conditioning", "visual learning", "linguistic learning (vocabulary, poems, etc.)", "concept formation", "problem solving", etc. (see Bredenkamp and Bredenkamp 1974, 610). However, when it comes to the question faced by the infrastructure planner—i.e. whether there are enough schools in a given region—such a differentiation is too finely grained and therefore irrelevant. It follows that the question of what should count as a mechanism is one of definitions, and must therefore almost always be answered relative to the planning issue at hand.

The choice of a given degree of resolution in the course of planning is always accompanied by a certain amount of focusing, and thus a narrowing of perspective. We cannot take into consideration *all* mechanisms that are theoretically possible. Consequently, we only gradually draw into view the mechanisms that are actually operative in reality. Those mechanisms which are included always form only a subset of all those that could be included. This, however, is not a weakness that could be avoided somehow—it is rather both unavoidable and necessary. Without the abstraction thereby enabled, planning would be impossible. When we, whether knowingly or unknowingly, accept the limits imposed by the specification of certain mechanisms, we often do so only provisionally. It is always possible to reject the choice that was made, restructure the whole thing, and thus transcend any given narrowing of perspective (see Winograd and Flores 1989).

Our knowledge about complex constellations of mechanisms can, with some simplification, be put into three categories: "black boxes", "gray boxes", and "translucent boxes" (see Bunge 1999b, 35).

- In the case of "black boxes", only what is externally observable is correlated—specifically, inputs and outputs. (Examples: the application of chaos theory in planning, every purely descriptive model of social processes such as migration, or any kind of data representing chronological sequences.)

- "Gray boxes" are partially "translucent". They include rough descriptions of how a system functions. In the case of gray boxes, so-called intervening variables are incorporated, without, however, describing the mechanism in detail. (Example: the theories of social mobility.)

- "Translucent boxes", on the other hand, contain detailed descriptions of the mechanisms at hand. (Example: population dynamics; mechanisms: deaths, births, moving to a place, moving away from a place.)[22]

Only translucent boxes deliver sufficiently detailed descriptions of mechanisms. In the case of black and gray boxes, we can only speculate about the mechanisms at work.

Our understanding of these relationships is further complicated by the fact that any change relevant to planning in all likelihood also includes biological, psychological, demographic, economic, political and cultural components, either simultaneously or sequentially. Consequently, we are almost always dealing with a combination of highly distinct mechanisms. Therefore, almost all single-factor (and especially mono-causal) descriptions are partial at best.

Mechanisms versus Forces

We distinguish between essential and non-essential mechanisms in a system: essential mechanisms are what constitute a system at its core—we call these simply "mechanisms". Most (essential) mechanisms are influenced by non-essential mechanisms, which we call "forces". Forces can be inhibited without changing the nature of the system. Thus, the mechanism "learning" can, for instance, be postulated as an (essential) mechanism of education in a school. However, other mechanisms that are also to be found in schools, such as coordination, internal power struggles,

---

22 In the case of "deaths" and "moving to" or "moving from", the relevant mechanisms can also be described in greater detail: scientific research deals with the mechanisms that make people age and die. Besides accidents, these include repeated damage and reordering of DNA or apoptosis (genetically programmed death).

Mechanisms involved in moving to or away from some place: changes in personal relationships, such as marriage, divorce etc., changes in the job market (layoffs, new hiring), choice of an educational institution (for example, a certain major) in a different city, or some combinations of these.

etc., are forces. Although forces are causally efficacious and thus significant, they do not define the nature of the system.

In other words, most mechanisms, be they physical, chemical, biological, social etc., are influenced by forces.[23] Forces alter the speed or the manner of operation of the system's mechanisms. Accordingly, a social force is an internal or external social factor that alters the speed or manner of a mechanism's operation in a social system. If a force is active within a system, it either accelerates or decelerates certain mechanisms of the system.

That is to say, the existence of a force implies the existence of a mechanism. Elections, public debates, and demonstrations are mechanisms of democratic change (or stagnation), but are not forces. In contrast, lobbyism, the manipulation of public opinion, bribes, violence, etc., are forces, because they alter the mechanisms of democratic politics.

It isn't always *social* forces that must be assumed in order to explain *social* changes. Natural disasters like floods or earthquakes can be regarded as forces that have social effects, but this does not yet make them social forces. That is to say, while every social mechanism has social effects by definition, not every social change results from a social mechanism.

While some forces act in the same direction, others work against each other— although not necessarily in a destructive manner. For example, the effectiveness of an organization depends on a balance between the forces of standardization and innovation, discipline and initiative, cooperation and competition. If a suitable balance is not reached, the system may stagnate or break down.

What counts as a force is—as with mechanisms—a matter of definition, which is to say a matter of one's point of view. After all, not every lobbyist will agree with the description of lobbyism as a "non-essential" mechanism.

In summary: forces form, mould, and shatter systems. They do this by altering the manner or speed of a mechanism's activity. Forces may be so strong that they force the agents involved to dismantle an old system and construct a new one.

Mechanisms versus Explanations

In our discussion thus far we have been using the concepts "explanation" and (essential and non-essential) "mechanism". The expression "mechanism" denotes what factually exists. In contrast, the expression "explanation" designates the *description* of the mechanism at the conceptual level.[24] This distinction between

---

23 Forces play only a negligible role or no role at all in certain technical processes, such as in spontaneous radioactivity, the diffusion of smoke in thin air (in which molecular collisions are less frequent), or in the propagation of electromagnetic waves in a vacuum.

24 Concerning the question "in what sense do mechanisms exist?", the following positions are among those available to us: most of the authors in the reader on social mechanisms released by Hedström and Swedberg (1998) define mechanisms as constructs, which is to say as mental fictions. In contrast, the view prevailing among scientists and engineers (one that we adopt as well) is different. According to this view, mechanisms are not just conclusions or judgments existing only in the "Cognitive Apparatus" of those imagining them, but components of the real world, just like objects and events—in other words, facts. (The existence of an object in

mechanisms and their respective explanation makes it clear that one and the same mechanism can be conceptually described, which is to say explained, in different ways. This distinction furthermore implies that the mechanisms operating on a factual level can diverge from the description or explanation given of these mechanisms. Consequently, explanations may or may not describe mechanisms accurately.

The distinction between mechanism and explanation also offers us the possibility of realizing that many a hypothesized mechanism may not exist in reality. Examples may include Adam Smith's "invisible hand" as described in his *Inquiry into the Nature and Causes of the Wealth of Nations*, the mechanism of divine providence, or the "urban compulsion" [*Stadtzwang*] postulated by Mönninger (Mönninger 1999, 10).[25]

With this in mind, Selle asks whether large-scale events like the Expo 2000 really have the effects that their supporters hope for,[26] or whether the effort expended by cities to host such events isn't rather more reminiscent of so-called "cargo cults". He, somewhat pointedly, compares the staging of such events to the behavior of natives in New Guinea that was (supposedly) observed by anthropologists. These anthropologists saw that "natives cleared large rectangular sections of jungle. They drove red and white poles into the ground, defoliated a tree and hung a fluttering bag in another. Afterwards, they took turns sitting at the edge of this field with their eyes directed skyward, and waited. For what? For "cargo". The scouts of the tribe had discovered that great birds descended on certain spots in the land to bring cargo (that is, precious goods) to the white man. These spots were staked out with red and white poles, antennas were placed at the edge (hence the defoliated tree), and a wind sock hung on one of the poles. The natives recreated all of this with the materials available to them, so that the gods may send gifts down to them as well... Several aspects of major projects in the nineteen nineties are reminiscent of such cargo cults. The native propagandists of big events sit by the edges of construction sites and wait for the external symbols of wealth to fall from the skies ... " (Selle 1994, 47)

The fact that we overestimate ourselves in planning and display an illusory optimism regarding the achievability of particular results and the effectiveness of the mechanisms we employ, is called an illusion of control, a well-known phenomenon in cognitive science (see e.g. Gollwitzer 1996, 559 ff). "It is possible that illusions of control are quite generally the product of states of consciousness related to planning." (Gollwitzer 1996, 562)

---

a particular state or the occurrence of an event in an object are called facts. "... a (real) fact is either the being of a thing in a given state, or an event occurring in a thing." (Bunge 1977, 267))

25 "Affluence and freedom of choice allow a much greater degree of independence from the urban compulsion (*Stadtzwang*)." (Mönninger 1999, 10)

26 We are working under the assumption that the Expo 2000 was not consciously planned as a way to modernize and expand the fairgrounds of Hannover at the expense of a third party from the very beginning.

Properties of Mechanisms

The following properties of mechanisms are of particular interest in the present context.

Mechanisms are Usually Concealed

Most mechanisms, whether physical or social, are concealed.[27] We therefore don't see the mechanisms, such as low investment rates, technological conservatism, or labor unrest, that cause a planning firm or a business to go under, just as we don't see the mechanisms governing the movement of the planets, telecommunication, or metabolic rate. Since mechanisms are usually concealed and thus not accessible simply by closer inspection, we can only gain access to them in one particular way: they must be expressed as explanations via language (spoken or written) or signs if they are to be intelligible to third parties.

Mechanisms are System-Specific

Mechanisms are system-specific, which is to say dependent on the object of investigation. There is no single mechanisms that occur, or are active, in every system. This is why mechanisms that are pertinent in physics or chemistry, for instance, are useless in planning. This also means that there are no universally valid mechanisms (see Popper 1987), such as those postulated by, e.g., Georg Wilhelm Friedrich Hegel and Karl Marx, as well by Herbert Spencer or Auguste Comte, to account for the development of societies (see Schimank 1996, 22).

However, even within a specialized discipline such as planning, caution is in order when it comes to the search for any extremely general mechanisms. Such broad questions as "why and how do cities evolve?" are occasionally still debated. This question, however, presupposes that all cities are more or less the same, and that more or less the same mechanisms bear upon all cities—which of course isn't the case. A particular city may develop because the local car company is currently achieving high profits and is investing in the area accordingly; another city because, being a transportation hub, it attracts companies that are dependent upon highly mobile consumers or workers; a third because it exists in the context of a federal system and is home to a great many administrative bodies and their numerous employees; a fourth because it profits from its location in the mountains by functioning as a tourist destination; a fifth because particularly wealthy people, demanding an accordingly exacting ambiance, have chosen to retire there—and so on. Without knowledge of the special mechanisms and agents involved, along with the specific circumstances, no correct answer can be given to the initial question "how and why do cities evolve?" More restricted questions like "why does a city of type A evolve in a situation of type B at a time C, etc.?" are therefore often more appropriate. Such questions presuppose that cities are only comparable in certain respects, not in all.

---

27 This is the case even for an old grandfather clock, because the gravitational field pulling at its weights is invisible.

Even though mechanisms are system-specific, it is nevertheless possible to group them according to their degree of similarity. Thus, competition between cities for companies in a particular line of business is similar to competition between two populations for a given resource. Such analogies cannot explain special processes. But they may nonetheless serve as heuristic devices in dealing with specific mechanisms by highlighting important aspects and thus facilitating the discovery of new mechanisms.

Mechanisms Only Exist in Concrete Systems

Since mechanisms only exist in *concrete* systems, it makes no sense to look for them in ideas or abstract objects, which is to say in *conceptual* systems, as in, for example, sets, functions or algorithms, because nothing happens of itself in such systems. The construct "mechanism" is therefore not valid for logic or mathematics. Similarly, there are no mechanisms in strategies, methods or plans: only in their implementation in physical or social situations do mechanisms play a role.

Methodological Tips Regarding the Development of Explanations

Mechanisms are by no means always easy to discover, that much has become clear from our discussion thus far. Specifically, there are other types of relations that may be mistaken for descriptions of mechanisms (i.e. for explanations). This is to say that not every representation that looks like a description of a mechanism at first glance really is one. In the following section, nine methodological tips that may aid in the development of explanations will therefore be offered.

Dynamic versus Kinematical Descriptions

A description of a process without any reference to the underlying mechanism is called kinematical. Kinematical representations of changes are purely descriptive and contain no explanations. In contrast, so-called dynamic representations make explicit reference to the mechanism in question.

Accordingly, while the observation that "city X is losing inhabitants for the first time since the middle ages" (analogous to Venturi 1998, 58) does describe a process, it contains no indication of what caused this process. If the mechanisms of a process are known, the (kinematical) descriptions of the situation can be derived; the reverse is not true: a description of a process does not involve the description of any mechanisms. Rather, one and the same description of a process can be explained by various mechanisms. If the description of a process is all that we know, we can only speculate about the various mechanisms that might underlie this process.

What goes for *single* descriptions of situations of course also goes if *many* such descriptions are aggregated into data.[28] Such descriptive statistical data, however,

---

28 Data are information about facts ascertained through well-defined methods of measurement. These methods of measurement rely, either implicitly or explicitly, on theoretical assumptions.

only unite a more or less large number of individual descriptions of states or processes, but don't identify any mechanisms. Mechanisms therefore generally cannot be inferred or deduced—in a stringent, strictly logical sense—from such data. Mechanisms must rather be conjectured. Thus, differences in the utilization of a particular traffic system can be represented through chronological data. For a planner, however, such data generally do not suffice. In order to be able to change the utilization, the planner must know which mechanisms created and stabilized the differences in utilization in the first place. Similarly, economists cannot infer mechanisms from economic indicators or chronological sequences. Such a causal connection must rather first be invented and subsequently tested.

The following text, for example, while describing several processes, does not give any information about what causes these processes:

> at the world's current stage of development, extensive spatial restructuring processes are taking place that, according to the opinion of many urban researchers, are bringing about a revolution in urban development. Among them are the displacement of industrial centers and centers of growth on a global scale, an accentuated differentiation of the types of urban development found in industrialized countries, the replacement of the relatively constant growth of urban agglomeration that persisted until the 1970s by a new form of development which is characterized by the splitting apart of cities into, on the one hand, stagnating or declining regions and, on the other hand, prosperous regions—a form of development that brings with it an increasing social and economic polarization. (Krätke 1995, 16; italics omitted).

Explanations versus Statistical Correlation

Statistical correlations are also mistaken for mechanisms. Two variables are said to be positively correlated if high values of one typically occur alongside high values of the other, and vice versa. As we all know, however, such a correlation does not mean that one variable is the cause of the other.

If one, for example, discovers a positive correlation between particular neighborhoods and a high crime rate within these neighborhoods, then this doesn't necessarily mean that the quality of the housing or environment is therefore the cause of the spatial distribution of criminal behavior (see e.g. Flade 1990, 518ff). Correlations do not explain mechanisms, but rather demand such explanations themselves.[29]

---

29 Another phenomenon belongs in this context as well: frequently, statistical correlations are initially taken to be inaccurate because no plausible mechanisms exist that could explain these statistical correlations. Stanley Jevon's hypothesis that sunspot cycles could explain economic cycles was, for example, not taken seriously at first because no one could imagine a mechanism that would connect sunspots to economic activity. Only recently has it been possible to show that sunspot cycles affect the terrestrial climate (and thereby agriculture) because sunspots are in fact strong electromagnetic storms that cause a rise in radiation and consequently an increase in the amount of solar energy reaching our planet, which in turn affects agriculture. (However: even this explanation does not yet suffice to explain the ups and downs of economic cycles.)

The concept of causal analysis can certainly be found in the social sciences (see Schroeder-Heister 1984, 371) where it serves in the analysis of statistical correlations between variables in which one variable is seen as the cause of another. Strictly speaking, however, there is no such thing, because upon closer inspection a given variable is not "explained" by another. One can only really say that a particular variable is a function of another variable (which is sometimes computable). That is to say, there is no "explanation" of variables by other variables, rather, one is merely analyzed with the help of the other, and this without reference to any mechanisms.

Explanations versus the Assignment to Classes

The explanation of a causal connection is also to be distinguished from the mere assignment of something particular to something more general, because no understanding of the mechanisms is achieved by means of such an assignment: all that occurs is that a particular fact is identified as being a member of a class. Unlike the mere assignment to a class, an explanation explicitly refers to a known or supposed mechanism.

Two examples: institutions that function according to the rules of civil service are not thought of as particularly efficient, customer-oriented etc. If someone now realizes that a company which regulates local traffic is to be assigned to the class of "civil services", then this does not yet explain the mechanisms that may cause the local traffic not to be efficient or customer-oriented.

Or: so-called global cities (New York, London, Tokyo etc.) register unusually high rates of population growth, economic growth, etc. If a city is assigned to the class of global cities, then this does not yet explain the mechanisms that cause these higher-than-average growth rates.

Explanation versus Teleological Explanation

Explanations also differ from teleological explanations. By teleology, we here understand the goal-directedness of a process. The actual or possible achievement of a particular state is then regarded as an explanation for our comprehension of a process moving towards this state.

In planning, it is certainly necessary to specify the goals that are to be achieved by a plan. Nevertheless, the question presents itself whether the comprehension of a state that is to be achieved really says anything about the mechanism involved—i.e., whether it really offers an explanation. Here too, the answer is 'no'. After all, we don't understand the mechanisms operative within a traffic system any better if we specify and pursue a particular goal, for example: "the number of commuters that use public transportation in the area are to be doubled". In order to alter systems according to our wishes, we need to know their mechanisms, not just goals.

Explanations versus Functional Explanations

A special case of teleological explanations are functional explanations. The following thesis contains a functional explanation: "character A evolved to serve function B, which is in turn necessary for the viability of the system". The presence of certain properties in a system is therefore explained by saying that these properties *fulfill functions* that are necessary to the system's normal operation.[30] Such functional explanations cannot, however, be equated with mechanisms because mechanisms bear no relation to particular adaptations, values or uses of a system, especially because the mechanisms of a system can also be insufficiently adapted.

Functional explanations are important in the social sciences, but they do not suffice in planning because they do not deliver the information about mechanisms that is needed for the kind of practical activity through which things and systems may be changed. An example: the thesis that the "function of the city is to be the central point for supplying a more or less large surrounding area" (Krätke 1995, 28) explains neither why cities emerge nor which mechanisms accomplish the supply of surrounding areas. Again: the underlying mechanisms must be discovered, especially because any given function can be fulfilled by *various* different mechanisms. Thus the function "supply surrounding areas" can be fulfilled by various mechanisms. The fact that no one-to-one correspondence exists between functions and mechanisms furthermore demonstrates the limitations of so-called functionalism, whether in the social sciences or in planning. A process should therefore be explained by mechanisms, and not by reference to its functional value.

Explanation versus "Comprehension"

The representations of mechanisms also differ from the interpretive explanations (*verstehen*) favored by hermeneutics. This latter approach assumes that social facts are only "comprehended" when they are "interpreted", which is to say, when the "sense" or "meaning" that this fact has for someone has been pinpointed. The process of comprehension that is involved has been variously characterized: as empathy by Dilthey, as the attribution of purpose by Weber, and as a reconstruction of the reasons that drive the agent by Pareto and Boudon (see Bunge 1999b, 19f). Such an understanding of the process of comprehension does not, however, establish any connection to mechanisms. It involves only an allusion to an inner, mental source of individual actions while disregarding external influences.[31]

Explanation versus "Narrative Explanation"

Athearn (1994) has suggested that we not look for mechanisms that are, in the context of a certain field of knowledge or activity, more or less generalizable and therefore transferable to analogous situations, but rather to replace these with "narrative causal

---

30 We are, for example, dealing with a functional explanation if the existence of gills in fishes is explained by stating that their function is to secure a supply of oxygen.

31 For details, see e.g. Bunge 1996, 150ff.

explanations"—i.e. with descriptions of particular cases. But these too do not suffice for planning. Were we to follow Athearn, we would receive only *ad hoc* descriptions, which is to say more or less plausible stories instead of the kind of generalizable and transferable explanations required in planning.

Explanations versus Tautological Descriptions[32]

Explanations are occasionally confused with tautological descriptions. Thus, for example, the thesis that cities develop along a "path of least resistance" is purely tautological. After all, such descriptions offer no explanations as to what mechanisms create this resistance. The same thing goes for a description which propounds that a development in traffic engineering has reached a "bottleneck". This description also does not answer the question about what caused this bottleneck.

Top-down and Bottom-up Explanations of Behavior Patterns

As has already been mentioned above, the task of planning consists not only in assigning locations and maintaining facilities (such as buildings, streets, parks, etc.), but also in directing behavior patterns—in for example introducing new rules regarding the use of facilities, something which is not a change in the facilities themselves, but rather in our interaction with them (see, e.g., Heidemann 1995 or Schönwandt 1999, 32f). That is to say that in planning we are also concerned with mechanisms that influence the behavior of people. We can find many examples of this in practice: there are not only streets, but also the rules governing their use (i.e. the rules of the road); there is restricted parking, car sharing, car pooling, toll fees, management of freight traffic, and more. All of these are approaches with corresponding mechanisms aimed at regulating and directing the patterns of behavior found in particular facilities. Accordingly, many campaigns in planning aim to alter the behavior of human beings via mechanisms—such as using public transportation instead of cars, sorting garbage, ventilating homes in an energy efficient manner, using less electricity, etc.

With this in mind, the following question presents itself: what influences human behavior?[33] Or: what are the mechanisms appropriate in this context? In this regard, two different perspectives—namely the bottom-up and the top-down approach—should not be considered in isolation, because we are always dealing with a mixture of the two. In the present context, bottom-up means that every individual activity also has structural results; every social fact is, after all, the result of individual actions.

32 By tautologies we do not mean pleonasms like "white mold" or "each and every", but circular definitions in which the definiendum appears in the definiens (see Lorenz 1996, 213f).

33 The question 'what influences human behavior?' is the starting point for one of the most encompassing and longstanding debates in the social sciences (psychology, sociology, economics, etc.; compare, for example, Heckhausen 1974, Bem and Allen 1974, Graumann 1975, Bem and Funder 1978, Giddens 1988, Esser 1993, etc.). This discussion cannot and will not be revisited in full here, rather, one of its main points will be summarized below.

The top-down approach, on the other hand, assumes that individual behavior patterns are influenced by social conditions. What is necessary is to take both perspectives into consideration simultaneously. Both viewpoints are, for example, integrated in the following hypothesis: "human conflicts primarily have two possible sources: interest in the same scarce resources or divergent goals within a social system." The individual choice to cooperate or to compete is certainly an individual thought process, but one which is at least in part induced by something external to the "Cognitive Apparatus", such as the lure of a resource or the protection or threat of a social system.

Explanations that ignore either of these two aspects are inadequate in the present context. An example of this is Schelling's (1978, 139) thesis regarding social segregation and concentration such as has emerged in many U.S. communities that are inhabited exclusively either by whites or by African Americans. Schelling holds the view that such segregation is primarily the result of individual choices guided by preferences: "choosing a neighborhood means choosing neighbors." Thus, for example, the choice of a neighborhood is equivalent to choosing a neighborhood of people who want good schools. Schelling ignores the fact that many people in the U.S. are forced to live in ghettos because they cannot financially afford to choose "good" neighborhoods or schools. His explanation overlooks the fact that individual intentions and expectations, and thus individual choices, are substantially shaped by social circumstances.

The fact that urban planning, on the other hand, often overemphasizes the top-down approach is shown by the following example: living and working, as divergent uses of space, cannot, as some planners proclaim, simply be recombined via a top-down approach.

> The debate over functional integration versus functional separation is carried out primarily by means of urban design. One tests … whether the separated functions may not once again be compatible with each other … This … perspective disregards the fact that the functional separation isn't exclusively the result of legally mandated emissions buffer-zones and the technical and functional requirements of the different types of use. The separation also emerged through numerous individual decisions on the part of users regarding location, and is an active exploitation of any special advantages of different locations. Households and companies only choose locations in close proximity to each other if this doesn't violate their interests and if this new location offers certain advantages. (Bonny 1998, 243)

The explanations developed in planning (i.e. descriptions of mechanisms) should do justice to both perspectives—namely, that we shape our environment just as our environment shapes us. Individual action and environment always go together because they mutually constitute each other. Consequently, individual actions are best understood if they are placed in a context, while this context is best understood if we analyze the individuals acting within it, along with their relations to each other.

In this section we discussed mechanisms, their significance for planning, as well as several mistakes that are possible when dealing with them. In the preceding sections, several points relating to the topic of constructs were discussed. In the following section we will explain how, or according to what steps, constructs are formed, and at which stages of our work in planning they may be found.

3.5   The Formation of Constructs

Concepts are defined using attributes. With the aid of relations, these concepts can then be combined with each other to form propositions. However, in order to provide an accurate and fitting representation of a system, a whole bundle of constructs (i.e., concepts, propositions in their respective contexts) is usually required. Whenever possible, these constructs are connected as a whole series of logically related propositions: a theory. We distinguish four stages in the process of construct formation: the list / the inventory, the sketch / the diagram, the specific theory / the theoretical model, and the generic theory (for the following description, see Bunge 1974a, 99f).

List or Inventory

The easiest way to form constructs or concepts is simply by listing material objects. Alternatively, one might also describe these objects, their constituent parts, and the general classes to which they belong. Should one wish to form a construct or a concept about a particular settlement's public transit system, then a deciding factor for this concept would, for example, be which users of the transit system will primarily be taken into consideration: whether it be the wheelchair-bound, cyclists, pedestrians, etc. The individual properties of each object are therefore grasped conceptually and presented as a list. In so doing, one obtains a bundle of unrelated information, which may end up serving as the basis for further deliberation.

However, lists such as these do not convey any information about the relationships that may exist between individual elements. We cannot discern if there is a relationship between the objects and attributes on the list and, if so, what kind of relationship it might be.

Sketch or Diagram

As a rule, simple lists of objects and attributes are therefore not enough to solve problems in planning. Constructs that must be formed in planning require that we elucidate the relationships that exist among the objects and attributes on this list.

In this second stage, we develop a sketch that displays not only the constituent parts of each object but also any relationships that may obtain among them. These connections are often represented as arrows in a diagram. In this way, we can show, for example, the different ways that products, money, etc., flow through a city, such as between the municipal government, employee households, employer households, and land owners (private and corporate landlords), etc. (see Krätke 1995, 38). Other examples of diagrams such as these include flow charts of a factory's production process; organizational charts of institutions; graphs for the maintenance and demolition of buildings, neighborhoods, or cities, etc.

Developing diagrams makes the formation of constructs more precise. The objects and attributes contained in the list are apprehended from a certain perspective and are put in relation to one another. Diagrams include all the information on a list, while they also clarify the various relations that exist between the individual components.

Specific Theory or Theoretical Model

The third stage of construct formation is the specific theory. A specific theory fills out the sketch. The relationships between objects and attributes represented in diagrams gives us only very limited information about the real connections. They only tell us which objects of an inventory are even part of a relationship or which objects stand in a certain relationship with one or more other objects. But they do not yet tell us anything about the kind of relationship that exists between the various components.

The third stage of construct formation consists of making the relationships represented in the diagrams explicit. In order to understand the represented connections and to make a construct intelligible, one must formulate a diagram linguistically. We must determine the kind of relationship that exists between each individual component—such as, whether we are dealing with a functional relationship or with a mechanism—as well as all the individual components themselves.

Generic Theory

The fourth stage of construct formation is the generic theory. While the specific theory, which is to say the theoretical model, fills the sketch out with additional information, the generic theory is free of all specificity. Rather than taking any specific objects into account, the generic theory holds true for all the objects that belong to a given category. The difference between the generic theory and the specific theory therefore lies in the objects to which it refers; while the specific theory deals with a more limited set of factual referents, the generic theory deals with a more general set. In the generic theory, the attributes of a group of objects are grasped in their connection to one another and described at a sufficiently high level of abstraction so as to apply to all the objects in a given category.

Logically, there is no difference between a specific theory of the third stage and a generic theory. During the fourth stage, the constructs are formulated in such a general way that they no longer refer only to individual objects. A generic theory such as Darwinian evolution, for example, applies to all living organisms and describes the creation and formation of every species.

The transitions from specific theories to generic theories are smooth. For our purposes, they depend mostly on how the specific planning question in the context of which the constructs are used has been framed.

However, urban planning does not have too many generic theories of this kind. After all, theories about "the organization and development of urban spaces" exist "only in rudimentary form… Even the various academic disciplines—from economics and sociology to geography—have not yet managed to generate or synthesize sufficiently comprehensive theoretical foundations." (Friedrich Ebert Foundation 2000)

To summarize: The previous sections dealt with four basic types of constructs. Then we described how concepts are formed. In addition, we brought up the topic of mechanisms. Finally, we provided a sketch of how constructs are formed.

This is not to say that, in actual practice, planners necessarily proceed as described when they generate constructs—which are bearers of knowledge—in order to comprehend a situation.

Rather, in order to comprehend a situation and to come up with proposals for how to act, planners tend to use so-called "schemas", "mental models", and "metaphors" as well as "analogies". The following section attempts, in rudimentary form at least, to elucidate these concepts.

3.6   Schemas, Mental Models, Metaphors and Analogies

Although it remains open to revision, a carefully thought-out construct can be seen as the "end-state" of a cognitive process (see Bunge 1996, 105). However, as we all know, in actual practice planners are often forced to make do with various "transitional stages." After all, not every construct or concept is explicitly constructed, such as it would be if we were to begin by first listing all of its attributes. Some of these "transitional stages" are the so-called schemas, mental models, as well as metaphors and analogies.

Schemas

A schema is a structured cluster of concepts that contains information and is already available to our "Cognitive Apparatus". It is used to represent objects, states of affairs, events, and processes (see Eysenck and Keane 1998, 262). Schemas are therefore structures of knowledge or clusters that entail assumptions and expectations about specific objects, states of affairs, etc. (see Zimbardo 1992, 623).

Take, for example, what we know about the construct "house". It contains countless concepts, such as "wall", "door", "window", "room", etc., with their respective attributes and corresponding propositions, including: "houses have rooms", "houses can be made of wood", "houses have roofs", and so forth. A whole bundle of concepts along with the propositions that result from linking those concepts together are therefore connected to the construct "house" and stored in our memory. That's because we "know" what a house is; we don't have to go through the process of forming this concept from scratch every time.

The idea of a schema as an innate structure helping us to perceive and interpret the world comes primarily from Immanuel Kant (1787/1963). In the years following 1930, F.C. Bartlett used the concept of a "schema" in his research at Cambridge University. Bartlett asked the question of how our memory and understanding of events are adjusted to the expectation that people have in connection to these events. He assumed that these expectations are present in the form of schemas and conducted a number of experiments that illustrate the influence of these expectations on our thinking and our memory. In so doing, he discovered that people "reconstruct" past events rather than remembering them precisely (see Bartlett 1932).

In his studies on developmental psychology, Piaget also availed himself of the idea of a schema. He used schemas to describe the changes in cognition experienced by children as they grow older. Schema-theories became increasingly prominent, especially in the 1970s, when several theories surfaced that were based on this idea: Schank's (1972) "conceptual dependency theory" used schemas primarily in order to describe relational constructs. Rumelhart and some others (Rumelhart 1975, Thorndyke 1977, Stein and Glenn 1979) suggested the concept of "story grammars" as schemas that form the basis of our understanding of narratives. Schank and Abelson (1977) coined the term "scripts" to characterize sequences of stereotypical actions that describe what people know about typical situations that come up in our daily lives. Rumelhart and Ortony (1977, as well as Rumelhart 1980) came up with a generic theory of schemas. And Marvin Minsky, with his 1975 concept of "frames", proposed a similar construct that functions in the domain of artistic intelligence (see Eysenck and Keane 1998, 262f).

Even this brief overview shows that the concept "schema" has shaped discussion, especially in the cognitive sciences, for several decades now.

The problem that comes up in connection to the use of schemas can be illustrated by a simple and oft-cited study by Brewer and Treyens (1981). This study demonstrates how schemas affect our memory as well as our decision-making process: Test subjects were taken into an office one at a time. They were told that this was the office of the scientist in charge of conducting the study and that they should please wait there for a moment. After about thirty seconds, the scientist overseeing the study took them from this office to a room next door. Then they were asked to describe everything they could remember about the office they had just left (which, of course, was really the site of the experiment!). A curious thing about this room is that it contained no books or binders—the things that one might expect to find in an office. Brewer and Treyens assumed that people's memory would be greatly influenced by the schema "office supplies", i.e., their pre-existing knowledge about "normal" offices. Accordingly, almost one third of the test subjects "remembered" having seen books and binders in the office (see Anderson 1989, 121f).

Memories are therefore greatly influenced by the schema that we may already have for a given concept or a given situation.[34] This experiment shows that schemas can, among other things, lead our perception astray.

Schemas usually contain no assumptions about mechanisms; these will be represented in so-called mental models.

Mental Models

The construct of mental models is in principle built on the concept of schemas. While schemas represent clusters of concepts and their (non-causal) relations, mental models depict our understanding of the mechanisms that determine a situation (see Eysenck and Keane 1998, 388). Mental models are therefore consciously or unconsciously

---

34 Dörner puts it as follows: "Someone once said: 'you should regard your memory as your biggest enemy!' As a general maxim this is of course false, but when dealing with macro-operators [actions in planning] it is worth taking to heart." (Dörner 1995, 305)

utilized assumptions about mechanisms, i.e. explanations (see above, Section 3.4). Since the 1970s scientists have explored how people use mental models when confronted with difficult situations, either in daily life or when solving problems. The concept of a "mental model" first came to the public's attention in 1983 with the publication of two papers that bore the same name—"Mental Models"—by Gentner and Stevens as well as Johnson-Laird. Despite the fact that not all authors define the expression "mental models" in nearly the same way (see, for example, Eysenck and Keane 1998, 388), we can nonetheless pick out some common properties in all of these various definitions:

- Mental models represent the understanding a person has of a system's causal connections. Sometimes, they are accompanied by visual, pictorial images.
- If a system is to be modified by the intervention of planners, then mental models are used to predict the behavior of the system and to suggest possible procedures.
- Sometimes people possess several mental models (which is to say different explanations of mechanisms) that compete to represent a given system, either in its entirety or in its parts.
- Mental models are fragmentary, unstable, capable of sudden change, and are formed *ad hoc* to meet specific demands. As so-called *ad hoc* rationalizations, they are sometimes used to give an account of our own actions.
- They have no empirical foundations and, for the most part, escape conceptual scrutiny as well.[35] What's more, many people persists in exhibiting "superstitious" behavioral patterns despite the fact that everyone knows these are not needed to reach the desired goals (see, for example, especially Eysenck and Keane 1998, 388 or Reason 1994).
- As mental models are not subjected to systematic scrutiny, they are vulnerable to unconscious and innate cognitive tendencies that underlie planning (see the cognitive traps described in Chapter 2 above).

Metaphors and Analogies

When planners do not have access to suitable constructs that, as bearers of knowledge, are needed to work through a planning task they can make do with various alternative means. One possibility—as described above—is first to list the attributes, then sketch the connections, and finally produce a theoretical model. Often, planners opt for another procedure: if the directly relevant constructs, which is to say knowledge, is missing in a situation, then constructs from other, related situations are carried over to the situation in question. This is done with the aid of metaphors and analogies. The use of metaphors and analogies in the planning process consists of provisionally

---

35 The following example illustrates why one is well-advised to subject one's mental models to close scrutiny: McCloskey experimentally demonstrated (1983) that many students of physics have inadequate mental models concerning even the simplest physical processes in Newtonian mechanics.

structuring an unclear or poorly defined state of affairs as if it were like another, better understood one.

But what are these metaphors and analogies? A metaphor is a figure of speech in which an expression that describes an object is transferred to another object that exhibits certain identical or at least somewhat similar properties (see Bußmann 1990, 484). Examples include: "Cigarettes are like time-bombs" (Klix 1992, 289). Or: "the construction industry has a foot of water under its keel." "This forest-covered valley is the city's green lung."

The given expression therefore contains two concepts. A modified meaning is engendered in the first concept by dint of comparing it to the second concept (which is usually intended to be symbolic). "The second concept is called the vehicle of the metaphor, the bearer-concept. The first concept is ... the target of the second. It is also called the topic- or target-concept of the metaphor" (Klix 1992, 289).[36]

In contrast to metaphors, which usually have a basic structure made up of two concepts and are therefore called "two-term-expressions" (Klix 1992, 295), analogies tend to contain four concepts ("four-term-structures" Klix 1992, 289). Example: "a talented poet of the state, so says the analogy, is like a comely swan constantly walking along a highway. The analogy is built on a sort of double-relation that becomes apparent upon examination: a swan that is capable of flight but whose movements are excessively constrained is like a poet with an excessively and voluntarily restricted repertoire of themes at his disposal. Or another example: a revolution takes pity on the destiny of individual citizens just like an avalanche takes pity on the houses it sweeps away. Both pairs of terms share a mercilessness in the released forces: revolutions to people as avalanches to houses." (Klix 1992, 295)

To put it more generally, in both metaphors and analogies the contents of constructs—and therefore knowledge—are carried over from one situation to another, always under the assumption that at least one property of both situations is identical or at least similar.

Examples of metaphors and analogies in planning include the following:[37]

"...hence, the city... can be seen as a machine, an 'ocean liner.'" (Ipsen 1998, 48)

"The city, and ever since the sixties the country as well, is getting to be a completely technical space—one which is created on the drawing board and must be maintained like a car." (Ipsen 1998, 49)

"If the city is once again to serve as a growth medium for desperately needed jobs in small and middle sized firms, then it must present itself as a turntable between the economy, science and the neighborhood." (Feldtkeller 1998, 275)

---

36 Translator's note: the topic- or target-concept is often also referred to as the "ground" of the metaphor.

37 Metaphors and analogies also exist in other domains, such as organizational theory: "Organizations are alternately described as anarchies (Cohen and March 1974), swings (Hedberg, Nystrom and Starbuck 1967), space stations (Weik 1977), garbage cans (Cohen, March and Olsen 1972), aboriginal tribes (Turner 1977), squid (Geertz 1973), market places (Georgiou 1973) and data processing plants (Borovits and Segev 1977)." (Weik 1985, 72)

"The metropolis as a horn of plenty, from whose surpluses the provinces benefit well into their deepest recesses, is an illusion." (Fritz-Haendeler 1998, 275)

"The city of Chemnitz is attempting—for the third time now—to reacquire its heart." (Dören 1998, 187)

"Especially this functional aspect [of the city of Lübeck as a trade and services center] is the deciding factor for the economic and political development of the Hanseatic Town ... [T]he old part of Lübeck [lies]—in an extended sense like a heart surrounded by water veins—at the geographic center of a much larger urban/provincial organism, for which it acts as a pacemaker." (Zahn 1998, 173f)

"...an oft-stated metaphor maintains that a strong city needs a strong heart, and that the traffic arteries of the city are like blood vessels in the body. This directs our attention to certain remedies for congestion: build higher volume routes to the center of the city." (Myers and Kitsuse 2000, 229)

"Among the many stimulating ideas thrown out by Wilbur Thompson, the incubator and filtering is one of the most interesting. The economic strength of the large metropolis is based on its capacity to innovate and to nurture new firms in new industries. The agglomeration economies of urban scale provide the right climate and environment for incubation to take place. Moreover, this process not only continuously regenerates the economy because the newer industries subsequently filter down into smaller cities and into other regions. Thus, the national metropolises play a crucial systemic role in the development of the economy as a whole." (Richardson 1978, 264; cited after Tank 1987, 17)

Using metaphors and analogies brings with it a number of advantages:[38]

• Metaphors and analogies are an obvious way to transfer knowledge. How else are we to learn from our experiences, if not by transferring the results arrived at in one situation to a new, similar situation?[39]
• Metaphors and analogies often provide a condensed description of a state of affairs, without having to describe everything in explicit detail. The metaphors implicitly contain these individual aspects and the reader (or listener) can tease them out for themselves. If we were to say that a planner "fights like a lion", then we would not only have come up with a concise description of his character, but also one that accepts future emendations and addenda, as

---

38 "The use of analogies has played a role in human creativity, invention, discovery, and scientific progress the value of which cannot be overestimated (s. Hesse, 1970). Thus, Mendelejv had the idea of organizing the periodic table of elements according to each element's atomic numbers and ability to bond with other elements by transferring the two-dimensional ordering of playing cards in a card game to the ordering of atoms (s. Sergejew, 1971)." (Dörner 1995, 310)

39 Thus, Bohr and Rutherford thought that an analogy could be drawn between an atom and the solar system. The solar system is a kind of enormous model for the atom, because the relationship between the sun and the planets corresponds to the relationship between an atom's nucleus and the orbiting electrons.

well as one that encourages the reader (or listener) to fill it out with further details. It is not necessary to say that he fought "valiantly", "energetically", "fearlessly", "aggressively", etc. As a result of this compactness, metaphors enable "the predication of a whole cluster of properties [here: attributes], that would have otherwise required whole lists of predications, in a single word." (Ortony 1975, 49; cited after Weik 1985, 72)

- Metaphors and analogies allow us to characterize things or situations even though we may not be able to describe them precisely with concepts,[40;41] in some cases we cannot find the right words, "in this kind of a bottle-neck situation we utilize metaphors in order to describe what words cannot describe literally [i.e. on their own]. ... In order to understand what can bring about this ineffability of language, we can reflect on the following comment, that ... [someone] uttered, upon ... having drunk [his] first glass of [carbonated] mineral water: "It tastes as if my foot has fallen asleep" (Weik 1985, 73).

- Metaphors and analogies often resemble actual, "lived experience more closely and are therefore emotionally, sensually, and cognitively livelier ... The 'foot that has fallen asleep' does not only convey something inexpressible, but it also evokes a spirited and vivid image, which simultaneously addresses several senses" (Weik 1985, 73f).

The drawbacks of metaphors and analogies are that they require the two situations in question to be identical, or at least very similar. This, however, is rarely the case. After all, metaphors and analogies treat heterogeneous objects as if they were identical, and precisely therein lies their fault (see Warburton 1996, 11). Metaphors and analogies aren't isomorphisms.

When two objects share an identical (or substantially similar) property, we often assume that they must therefore also be identical (or substantially similar) in terms of their other properties, even if the latter cannot be observed directly. ("When the knowledge is transferred from one domain to another, there is a tendency to transfer

---

40 "At the moment, a very successful metaphor likens the layout of cities with various preparations of eggs: up until the end of the Ancien Regime, the city was like a hard-boiled egg with walls, which, like an eggshell, encompassed a compact and dense center made up of representative institutions, residential, and commercial buildings. Until the Second World War, the city was like a fried egg, with the yoke representing the old center of the city right in the center of the expanding and gradually thinning egg-white, which represented a city's outlying settlements characteristic of the industrial era. In the past fifty years we have been dealing with scrambled eggs, in which urban geographers assiduously busy themselves by— with the help of fractal geometry—trying to determine the form of each piece of ham, as well as its location relative to every other piece of ham, within the whole scrambled mess." (Venturi 1998, 66)

41 "A significant theoretical system in psychology, namely Freudian psychoanalysis, can to a large extent be seen as a transfer of the idea of energy in thermodynamics (s. Wyss, 1970, S. 30ff., S. 49f.). The soul is an entity that generates energy within the bubbling cauldron of our drives, which must then be 'freed' or 'discharged.' The ego, or rather the superego, directs the steam into the 'correct' pipes. If the energy 'cannot be discharged,' then the cauldron cracks and we have a 'neurotic symptom.'" (Dörner 1995, 310)

coherent, integrated pieces of knowledge rather than fragmentary pieces". Eysenck and Keane 1998, 395) An obvious example: We notice that a certain aspect of one situation corresponds to a certain aspect of another situation: the objects in our solar system attract each other just as the objects in an atom attract each other. Certain other properties of the solar system (planets "orbit" around the sun) have therefore been transferred to the atom (electrons "orbit" around the nucleus), despite the fact that nobody observed this behavior in electrons. As useful as this kind of transfer can be, the general problem in doing so remains: the similarity of two objects in one respect cannot guarantee that they are also similar in some other respect. At most, this principle can give rise to probabilistic inferences, but never definitive, reliable conclusions.

Another drawback of metaphors and analogies is that a description of a state of affairs that was arrived at with the help of a metaphor or an analogy is more resistant to modification than a description arrived at with the aid of predication. Apparently, we humans have a hard time modifying or adapting the constructs transferred by way of metaphors or analogies, because we have a tendency to transfer "coherent, integrated pieces of knowledge rather than fragmentary pieces" (Eysenck and Keane 1998, 395; see above) to a given situation (compare this especially to Dörner 1995, 314f or Pylyshyn 1986). A description of some situation arrived at through predication

> consists of linguistic markers that are each assigned to the components of a state of affairs as well as to the relations that exist between the components of this state of affairs. This allows for an easy manipulation of the description of some state of affairs. By changing a single piece of the puzzle, the entire picture of a state of affairs has been altered. To make the point, one could say that propositional codifications are *more flexible* than analogies. They are easier to modify. This turns out to be very important for our thinking and problem solving. After all, thinking and problem solving consist of producing something novel... It is easy to see the modification of the picture of a state of affairs as first grasping the state of affairs linguistically and then changing that linguistic understanding. ... The analytical character of language, the fact that language consists of individual words for individual objects and relations, is what allows my thinking to supersede reality. ... This sort of flexible treatment of states of affairs is only made possible by our access to language and it is the inestimable advantage of propositional codification. A purely pictorial imagination is always conservative, and cannot escape the domain of the initial experience. (Dörner 1995, 315f; italics in original)

In the preceding section we explained the steps that are involved in the formation of constructs. This section dealt with schemas, mental models, metaphors, and analogies. So far, we have not yet described when constructs have a "meaning", when they have both substance and content. The following section is dedicated to this topic.

3.7   The Meaning of Constructs

Constructs only have substance and content when they have a "meaning". However, in everyday life, the word "meaning" is used in various different ways: "The word *meaning* is one of the most abused in both ordinary language and social science. Pop philosophy speaks of the meaning of life, whereas exact philosophy assigns meaning only to constructs and their symbols, so life is neither meaningful nor meaningless. In social studies, too, there is careless talk about meaning of an action, referring to either the goal or the effectiveness of the action. I will steer clear of such equivocations, admitting only constructs as bearers of meaning." (Bunge 1996, 55)

According to Bunge, the meaning of a construct is defined primarily with the aid of the following constructs: purport, intension, import, reference, and extension. If a construct is to have a meaning, if it is to be more than just an empty phrase, then these five components must be described with adequate precision. (For the following elucidations, see especially Bunge 1974a and 1996.) These five concepts give us an idea of whether or not a construct has meaning; after all, we use expressions—such as "sustainability"—whose meaning is not always entirely clear all the time, not just in planning (see Rudolph 2000).

What do these five concepts mean?

Purport (Precursor Constructs)

By "purport", we shall mean all those constructs on which another construct's intension is built (the core construct, see below); all those constructs which are used to define that other construct's intension. They are the determinants of a construct. "… the purport of a construct in a given context is the collection of constructs upon which it depends, or which determine it…" (Bunge 1974a, 142).

Constructs are defined with the help of other constructs. For this reason, the process of defining a construct is in principle never complete. If someone were to attempt a complete definition of some construct, then—as we have already mentioned—at some point he would have to define electrons, neutrons, quarks, or even smaller particles. With this in mind, we term those constructs the 'purport' of another construct that, although we use them to describe the intension of the latter construct, we do not define "en detail". "Attempt not … to define *every* concept! It is impossible." (Vollmer 1993, 136; italics in original)

The following citation provides an example of how the concept "city" can be defined by other concepts: "[Hoffmann-Axthelm] defines city chiefly through the concepts 'immigration' and 'ecology.'" (Mönninger 1999, 22) Hoffmann-Axthelm has therefore made a choice that could prompt us to ask why, for example, economic or politico-administrative topics are left out of his definition. However, it does not follow that his choice was "wrong." The reason: "Definitions are stipulations, or conventions, not assumptions. They are true by convention, not by proof or by virtue of empirical evidence." (Bunge 1996, 69).

In the final analysis, the choice of a definition should be made such that it is appropriate to the treatment of the problem at hand.

Intension

A construct's intension is the collection of those concepts and propositions (in their respective context) which constitute the core of the construct that is to be described:

"... the set of constructs it subsumes or embraces—it's *intension*..." (Bunge 1974a, 116).

Some rudimentary examples:

A possible intension of the construct "sustainable resource management": "The rate at which a resource is consumed ought not to exceed the rate at which it can be regenerated." (see Schäfer and Schön 2000, 25)

A possible intension of the construct "pedestrian zone": "Pedestrian zone" is, as described above, a construct from the 1960s. The expression "pedestrian zone" refers to portions of a city that at one time were used to accommodate individual vehicle traffic (i.e. streets), but that today are only accessible on foot. The following exceptions are permitted: motorized vehicles (delivery trucks at designated times, ambulances, police vehicles, and fire trucks, sanitation vehicles); non-motorized traffic (skateboarders, inline skaters, the wheelchair-bound), etc.

A description of the intension of the "export-base-concept":

> The thesis that a city's income and therefore its development are determined by the income or capital generated from exports stands at the center of the export-base-concept... The development of a city's economy is, according to this concept, dependent primarily on the growth of an export sector spurred on by increased demand. The growth of a city's export market then stimulates that region's service sector, which, by way of a multiplier-effect, causes the city's entire economy to grow (e.g., in the form of city-wide increased employment opportunities). This is how the export-sector can be seen as the basis for a city's economy ('base-sector'), whereas those economic sectors that do not export anything are assigned a "secondary", which is to say subservient, function ('service-sector'). (Krätke 1995, 41)

These examples are here reproduced without their contexts in planning (i.e. the definition of a planning-task, the approach used in planning, etc.), and are therefore of necessity fragmentary.

In spatial planning, so far as intension is concerned, three points are especially noteworthy.

First: In order to be relevant for planning, the intension should contain a description of the given mechanism. This point was discussed in detail in Section 3.4.

Second: In spatial planning (architecture, urban construction, urban planning, regional planning, etc.), spaces and objects play a leading role. Spatial planning is "... a process of permanently having to deal with spatial problems as they arise." (Lendi 1998, 25) After all, dealing with problems in spatial planning almost always means elaborating instructions to modify a constellation of objects seen to be wanting. It follows from this that the intension must contain a description of those objects (in what follows, they will also be called "factual referents"). If the intension were to lack a description of the relevant factual referents, then it would lack precisely that with which planners must work. The construct would therefore have no bearing on the task of the planner and would therefore usually be irrelevant for spatial planning.

Strictly speaking, we can distinguish between three different types of constructs here: In the present context, constructs are *directly* relevant to spatial planning if they contain those factual referents with which planners must work. On the other hand, constructs may be *indirectly* relevant if, even though they do not contain any descriptions of factual referents, they stand in a conceptual relationship to some other construct which is itself directly relevant. If both are absent—both a description of the corresponding factual referents and a conceptual relationship to a directly relevant construct—then the construct is irrelevant to spatial planning.

Third: In addition, one comes across the following situations in planning, which are significant for interdisciplinary work: Assume that a planner and a social scientist are together working on the topic of "dwelling". For the planner, the following attributes of "dwellings" are, in the present context, especially significant: the rooms, in which people dwell, inclusive the walls, which enclose these rooms, the size of the rooms, the distribution of each room relative to the other rooms as well as the four cardinal directions, how each room is outfitted, the plumbing and wiring (electricity, water, sewage, etc.), and much more. The social scientist might define "dwelling" completely differently though: "It describes the physical, social, and psychological transactions by which a person maintains his or her own life, joins that life with others, creates new lives and social categories, and gives meaning to the process, thus gaining a sense of identity and place in the world" (Saegert 1985, 288). "This *place in the world* does not refer to a concrete place such as a dwelling in terms of the *sum of its rooms, which are made possible by managing the household* (Flade 1996, 485)." (Hellbrück and Fischer 1999, 387; italics in original)

Both descriptions use (on the level of sign/language) the same word, namely "dwelling". The intension of both descriptions is completely different though, because both enlist completely different attributes to make the construct more precise. So far as intension is concerned, then, the two constructs do not intersect with one another. Even though both participants use the same word in their discussion of "dwellings", they are nevertheless talking about two completely different things. Productive, interdisciplinary teamwork is not possible in these circumstances. If the social scientist wanted to support the planner in the latter's work, he would have to integrate those attributes with which the planner describes the intension of "dwelling" into his social scientific constructs while setting them in relation to the attributes of his own, social scientific description.[42]

Likewise, constructs from other specialized disciplines (geography, geology, meteorology, economics, etc.) often cannot be used for planning, even though they may very well be identical to the constructs of planners in terms of the signs/ languages that are used. Only an assimilation of the given intension ensures that everyone involved is in fact talking about the same thing. Consequently, planners sometimes have no choice but to work out new constructs if they are going to work together with members of other specialized disciplines—in practice, this is usually a thorny undertaking because even members of technical disciplines tend not to be familiar with the ontological and semiotic dimensions of constructs.

---

42 According to the author's own experiences of cooperation between planners and social scientists, precisely this is the main conceptual obstacle responsible for the fact that fruitful results remain the exception, not the rule, in attempts to get people from both disciplines to work together over the past thirty years (see, for example, Schönwandt 1982).

Import (Implications)

"... the import of a construct in a given context is the collection of constructs that hang from it, or that are determined by it..." (Bunge 1974a, 142). A construct's import therefore refers to those constructs that describe possible results, consequences, or ramifications of the former. It is therefore a collection of those constructs that, for their part, are determined by the intension.

"Sustainable development", for example, is defined as, among other things, the following: "Sustainable development has been accomplished when the people living today behave in such a way that the capacity of future generations to satisfy their own needs and to choose their own lifestyle is not jeopardized." Because this description focuses primarily on coming generations and their lifestyles, it concentrates on a construct's import, and hence the results of that construct. It basically has nothing concrete to say about how people who are alive today ought to act, and hence it does not concentrate on the construct's intension.

Reference

We are here dealing with the facts (objects and events) to which a construct refers: "... every well-formed proposition is about, or refers to, something or other. For example, 'flows' is about some fluid. The collection of references of a ... proposition is called its reference class. For example, 'mass' refers to all bodies ..." (Bunge 1999a, 244f). "The semantic concept of reference comes up when asking *what a statement is about* quite apart from the way it is conceived, applied, misapplied, or tested." (Bunge 1974a, 32)

We already noted above that intension is only directly relevant to spatial planning if it comes with a description of the relevant factual referents. If the intension is missing a description of these factual referents, then the construct is irrelevant in terms of how it bears on the definition of a planning task, and it is therefore irrelevant to all spatial planning.

An intension without a description of the factual referents represents one extreme. However, in urban planning, for example, we also find the other extreme: work, in which an attempt is made to describe a construct—such as "city" or "models of cities" (Raith 2000, 10)—only with the aid of factual referents, without simultaneously elucidating the given conceptual content of the construct. In his otherwise readable book, Raith falls into this trap, which is reproduced as Figure 3.6 (see Raith 2000, 10).

These kinds of representations focus on the spatial and material features of a city, and fail to be meaningful descriptions because they do not contain sufficient information regarding purport, intension, or import. As such they define the construct "city" about as little as the construct "state" can be defined through a mere description of a state's spatial/material territory.

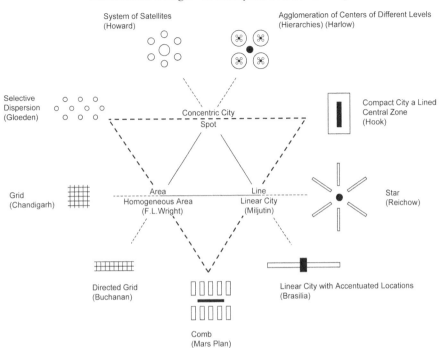

Figure 3.6   A system of possible models of cities by Gerd Albers

Source: Raith 2000, 10.

Extension

The difference between a construct's intension and extension was worked out by Gottlob Frege (1952). "The *intension* of a concept consists of the set of attributes that define what it is to be a member of the concept, and the *extension* is the set of entities that are members of the concept." (Eysenck and Keane 1998, 235) "Every predicate [=attribute] determines a class called the 'extension' of the predicate. This is the collection of individuals (or couples, triples, etc.) that happen to possess the property designated by the predicate concerned." (Bunge 1996, 52)[43] The extension is thus the set of all objects of which a concept is true, its scope. So, for example, the intension of the concept "bachelor" is the set of all attributes that define this concept (viz. male, single, adult), whereas the extension of this concept is the set of all bachelors in this world, from the Pope down to Mr. Schmidt next door. Note that the reference class of this concept is the set of all men in this world. Two further examples: The reference class of the concept "energy" is the set of all material objects, and the reference class of "poverty" is the human species, *Homo sapiens*. Since all material

---

43 "…while both predicates [=attributes] and propositions can be assigned referents, only predicates are normally said to have an extension. And even in the case of predicates a distinction between extension and reference class is needed." (Bunge 1974a, 119)

objects use or exchange energy, the extension of "energy" is identical to its reference class. However, since not all people are poor, the extension of "poverty" is only a subset of that concept's reference class.

A passage from Maurer (1998, 78) well illustrates the five concepts, purport, intension, import, reference, and extension:

> Part and parcel of the new 'Bahn 2000' initiative[44] is the renewal and development of regions adjacent to a railway station, which are mainly located in the middle or core of a city or region. Furthermore, connected to the initiative is the structuring of spaces whose density and construction make it possible to administer the trains as well as other, connected modes of transportation in a way that is both consumer-friendly and that only requires the input of a manageable amount of effort. This structuring of space requires that extant residential areas sprawl either not at all or at least not too far. It must also satisfy new needs as they arise in areas that have already been settled. As such, the initiative both enhances the economic potential of infrastructure and minimizes negative effects on the environment, such as, for example, the primary use of energy per person, the degree of soil sealing, and the degradation of the landscape. Likewise, the negative effects that can be expected to result from an overburdened road network will be mitigated, because alternative modes of transportation are available. (Maurer 1998, 78)

Examples of purport include: the concept of "Bahn 2000", regions adjacent to railway stations, city core, regional core, structuring of space, consumer-friendly, manageable effort, etc.

The intension contained in the citation can, in simplified form, be formulated as follows: "The 'Bahn 2000' ought to be administered in a way that is both consumer-friendly and that requires only a manageable effort. In order to achieve these goals, it is necessary (though perhaps not yet sufficient) to do the following: the areas adjacent to railway stations in the core of a region must be renewed, settlements cannot be allowed to sprawl too far from the regional core, and the desires for new construction must be satisfied within the extant settlements."

This intension contains, as a subset, the following mechanisms: The concept of a "Bahn 2000", together with the density of the settlements near the railway stations helps to bring about

- the enhancement of the railway system's economic potential
- a minimization of negative effects on the environment (primary use of energy, soil sealing, degradation of the landscape),
- a reduction of the traffic on streets, etc.

In addition, this intension carries with it certain explicit or implicit specifications about what should be done with respect to the factual referents, such as: "The settled regions ought not be expanded." "New apartment buildings, office buildings, etc., should be built within an extant settlement, not outside of it."

---

44 Translator's Note: 'Bahn 2000' is a nation-wide initiative to modernize Switzerland's railway system, with the aim to make public transit in that country run more smoothly and efficiently.

The substantial aspects of the *import* were already described (in the previous example) in the context of mechanisms. If Maurer's suggestions were implemented, in his opinion this would lead to the following consequences:

- the enhancement of the railway system's economic potential
- a minimization of negative effects on the environment (primary use of energy, sealing of topsoil, degradation of the landscape),
- a reduction of the traffic on streets, etc.

The factual referents in Maurer's description are, for example: railway buildings, apartment buildings, factories, streets, railway cars, oil, gas, earth, plants, animals, etc.

The extension in this example contains all the relevant/affected objects (railway buildings, apartment buildings, factories, streets, railway cars, oil, gas, earth, plants, animals, etc.) that arise in the railway stations newly managed according to the principles of the "Bahn 2000".

The upshot: To make working with meaningful constructs possible, the purport, intension, and import of a construct should be formulated with maximum clarity. Reference as well as extension ought also to be made as explicit as possible. Failure to do so means working with unclear constructs (see Section 3.11).

Up to now, we have described individual components of the semiotic triangle:

- language/sign
- objects/events and—somewhat more elaborately—
- constructs.

We shall now turn to the relationships that exist between these three components.

3.8 The Relationships Within the Semiotic Triangle

Initially, there are three relations in the semiotic triangle, namely those of denotation, designation, and reference—the "ways towards" that exist between the components, so to speak.[45] In addition, there are two "ways back": evidence and semiotic interpretation (for details, see Bunge 1974a, 1974b; although we simplify the relationships here).

---

45 Here we are dealing with the same relation—reference—as we were in section 3.7, "The Meaning of Constructs."

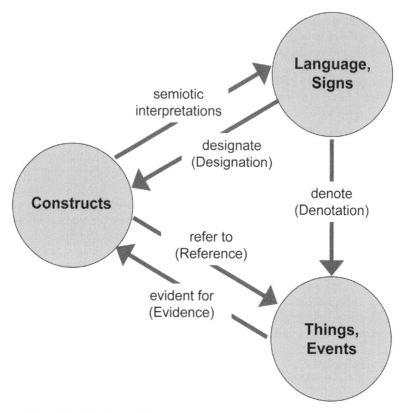

Figure 3.7   The Semiotic Triangle (expanded)
Note: For further explanation, see below

Denotation: Language Denotes Objects and Events

Language/signs serve a dual function in the semiotic triangle. The words that we use (among other things) denote objects and events. So far, we have said nothing other than that we can name objects and events, that we can "baptize" them. Certain words are assigned to certain objects or events. With this, language undertakes to secure the connection between objects/events and signs.

In planning, this connection between signs on the one hand and objects/ events or facts on the other is decisive for making plans clear and unambiguous (planning land use, or resources, etc.). This connection is indispensable for efficient communication.

In the spirit of this relationship (denotation), it is often a misunderstanding to say, "S *are* the participants in street traffic." Much clearer would be, "S denotes the (factual) participants in street traffic."

Designation: Language/Signs Designate Constructs

The second direction in which language signifies is towards constructs. We use many different constructs in our "Cognitive Apparatus" that are differentiated according to what they designate. When a construct is designated linguistically, such as, e.g., through the words "pedestrian zone" or "house", a specific construct is generated or invoked in our "Cognitive Apparatus". In this way, language activates constructs we have already learned and tested and thereby makes the knowledge contained in our "Cognitive Apparatus" accessible.

Misunderstandings can therefore be avoided by saying "S designates (the construct of) the participants in street traffic" instead of "S *are* participants in street traffic."

Signer provides a further example of how a linguistic expression or sign alone does not always make it obvious if it designates a construct or denotes an object. The sign is polysemic:
  "IRIS

- a women's name
- messenger of the gods (Greek mythology)
- a flower
- a muscular diaphragm in the eye

Acronyms:

- Infra-Red-Information-System (in connection with the automatic influence on traffic signals)
- Integrated Road Safety, Information and Navigation System (European research program DRIVE)
- Institut de Recherche et d'Information socio-économique, Travail et Société (Université Paris Dauphine)." (Signer 1994, 107)

Without further specification of a context, it is impossible to decide what the linguistic expression "IRIS" means here. If it is a name, then we are here dealing with denotation; if it's a construct, then we are dealing with designation.

Reference: Constructs Refer to Objects/Events

The construct "house" refers to the factual house. The construct "house" therefore has a counterpart in reality, namely the real object to which it refers: "...a (real) fact is either the being of a thing in a given state, or an event occurring in a thing" (Bunge 1977, 267; see Section 3.1). Example: The material objects of the construct "degree of motorization", meaning its factual referents, are—as described above— the inhabitants of a particular region as well as the automobiles permitted in this region.

The reference-relation therefore relates constructs on the one hand (concepts, propositions, contexts, theories) and material objects or events (facts) on the other hand. The class of material objects or events, meaning the referents of a given construct, are called "reference class."

More concisely, the path is as follows: from a linguistic expression → by way of designation → to the constructs, and from there → by way of reference to the objects.

While denotation, designation, and reference make up the way towards the objects/events, in what follows, we shall sketch the way back.

Evidence: Some Objects/Events Evidence Constructs

The relation of factual reference points from constructs to objects/events, which is to say facts (=reference class). The relation of evidence, on the other hand, points from objects/events, i.e., facts, back to the constructs; although only to that subset of constructs that are testable. Untestable constructs do not relate to any corresponding material objects or events (referents). That's because often only a few of a construct's assertions are even accessible for empirical testing.

This is to say that not all objects and events which can be invoked as evidence for a construct must in this sense also actually evidence something. The set of objects and events to which a construct refers is not automatically identical with the set of objects and events that evidence a construct. When discussing questions about the development of a city, one could, for example, examine the shape of a city's boundaries with the aid of fractal geometry. The question remains, however, whether we can read something out of the shape of a city's boundaries that refers to those relationships that actually influence the development of a city.

This means that reference classes are not evidence classes.

Semiotic Interpretation

Semiotic interpretation is the assignment of constructs (concepts, propositions, contexts, theories) to language/signs—though actually only to symbols and indices, since icons, while denoting objects due to their factual similarity, do not designate constructs. Semiotic interpretations are therefore the "way back" from constructs to language, meaning to the kinds of signs called "symbol" and "index". To put it differently: When we apply what we have said so far and make the meaning of constructs that make up a planning task more precise, then we are carrying out a semiotic interpretation. In so doing, we use mostly symbols—the linguistic expression "city development" is an example of such a symbol—and symbols are signs that require semiotic interpretation as a matter of principle. After all, the meaning of symbols rests, by definition, on prior agreements.

Semiotic interpretation is therefore required to clarify the assumptions that are contained in constructs and the objects/events linked to those constructs. This means that a large portion of the planner's job consists of semiotic interpretation.

We distinguish three kinds of semiotic interpretation (see Bunge 1974b, 1ff):

- Conceptual interpretation; here we are trying to clarify the relationship between two constructs, for example, the relationship between a generic construct on the one hand and a more specialized, "limited" construct on the other. Conceptual interpretations are construct-construct-relationships.

- Factual interpretation; here we are dealing with the relationship between a construct and the factual objects/events to which the construct refers; these, in turn, can also be evidence for the construct.
- Empirical interpretation; here we are dealing with the relationship between a construct and data[46] about factual objects and events that we may have at our disposal. Empirical interpretation deals with processed, calculated data, while factual interpretation deals directly with factual objects and events (i.e., without the intermediary steps needed to generate data).

An example: If, in planning a civil registry office, one must deal with a number, which is to say a set, of married couples, then the semiotic interpretation (see Table 3.2) goes—via language/sign—from the generic mathematical construct of a "set" (by way of conceptual interpretation) to a more particular construct such as "the set of all the couples", and from there (by way of factual interpretation) to the factual objects, such as a collection of the married couples or (by way of empirical interpretation) to the empirical data such as the collection of married couples as they were counted by a government agency.

Table 3.2   Kinds of semiotic interpretation (analogue to Bunge 1974b, 2)

Kinds of Interpretation	Relationship	Example
1) conceptual	generic construct --> particular construct	(mathematical) set --> set of couples
2) factual	particular construct --> factual object	set of couples --> collection of married couples
3) empirical	particular construct --> empirical data	collection of married couples --> counted collection of married couples

It is vital that interpretations not introduce any inconsistencies. They should be exact, not pictorial, and if possible they should not contain any analogies. Furthermore, in terms of the planning task they should be as comprehensive and universal as possible, not "spotty". This means that they should not only be valid in small sub-fields. Constructs are therefore symbolic, but not metaphorical or imagistic, even though they may contain some pictorial components.

To summarize: Even in planning, symbols symbolize nothing without semiotic interpretation. A symbol, such as the linguistic sign "city development", is totally lacking in content without semiotic interpretation.

---

46 "... in ordinary language 'datum' and 'fact' are often used interchangeably. This usage is incorrect, for data are propositions, not facts." (Bunge 1996, 85)

Up to now, we have described the components of the semiotic triangle. This chapter was dedicated to the relationships that exist between these components. In so doing, we kept on alluding to individual attributes of constructs. In the following section, we should therefore summarize the most important attributes of constructs. In Section 3.10, we will name several consequences that arise for planning. Section 3.11 deals with some possible points at which mistakes might creep into our work with constructs.

3.9 Attributes of Constructs

Constructs chiefly have the following attributes, which are of importance to planning.

Constructs are Bearers of our Knowledge

Surely the most important attribute of constructs is that they are the bearers of our knowledge—an aspect whose importance can hardly be overestimated, as it is this that affords them their prominence. Constructs are what bear a "meaning", "content", or "sense". "Meaning", "content", and "sense" are to be found neither in, by, nor on objects/events or language/signs. It is by dealing with constructs, then, that we deal with the real content, the conceptual core, of a topic. Without constructs, language/signs and objects/events are without content and thus without meaning—nothing but empty sounds and therefore "senseless" in the truest sense of the word.

Comparisons Between the Attributes of Constructs and the Properties of Material Objects

Other, at first sight less "weighty", attributes of constructs can be made more distinct by comparing them to properties of material objects. Material objects, whether alive or dead, natural or artificial, exhibit, among other things, the following properties: they are in a certain place, they have extension and they persist through time, they have energy, and they are capable of change (see Bunge 1983a, 47). In contrast, the natural number 7, the construct "region" or the construct "waste" (as opposed to the empty, dented aluminum can) are not in any specific place. These constructs have neither a spatial extension nor a temporal duration. Consequently, it is impossible to get them wet or to paint them. They have no energy. They are not in any particular state. Their state-space is empty, which is to say that constructs are not hungry or tired, like people sometimes are, they are not green or covered in moss, as are moist rocks. Because they are not in any state, they cannot change. This means, for example, when we speak of a "change" in a construct in the context of this book, we always mean "new" or "different" constructs. Constructs also cannot act, which means that they cannot sign a contract, cannot speak, cannot perceive, etc.

Constructs are Fictions

Whereas material objects really exist, constructs are cognitive artifacts, fictions.[47] They exist conceptually, on the basis of convention: "... constructs, they are total fictions: what is real is the brain process that consists in thinking of some object." (Bunge 1974a, 27) Constructs, then, are mental images, in principle nothing other than Asterix[48] or unicorns. As artifacts, they—being mental images—depend entirely on humans. If humanity were suddenly to die out one day, there would no longer be any constructs: no more "pedestrian zones", no "regions", no "waste", no "urbanity", and no more number "7". Perhaps there would remain some ink blots of a certain shape on pieces of paper, but there would not be anybody left who could read them and generate the corresponding construct in their "Cognitive Apparatus."

The mode of existence exhibited by constructs is one of the central themes of this book. As such, we will allow ourselves an excursus at this point.

Excursus on the Question: In What Way do Constructs Exist?

This question—'What is the nature of constructs?' and 'In what way do they exist?'—has fascinated and moved nearly every philosopher since antiquity. The main philosophical theories on this subject are well known. Bunge describes them as follows: 1983a, 42ff):

- Platonism. Conceptual objects (constructs) are ideal entities that exist as such (perhaps in Popper's "World Three"), independent of the physical world and especially the thinking being.
- Nominalism. Conceptual objects (constructs) are a subset of all the linguistic objects. They are signs and exist only as signs.
- Empiricism. Conceptual objects (constructs) are mental objects and exist in the same way as any other ideas, meaning as perceptions or images.

The following must be made perfectly clear for "naive realists" (see Bunge 1996, 354): None of these philosophical points of view assume that constructs are anything like material objects. They are either relegated to a separate, ideal world, or they are purely linguistic or mental objects.

However, none of these philosophical traditions does justice to the nature of constructs.

---

47 This is the main proposition of so-called epistemological constructivism, as it is, among other things, propounded by Aristotle, Kant, Engels, Einstein, Piaget, Popper, and Bunge. It claims that all concepts, theories, etc. are human constructions. This approach is not to be confused with ontological constructivism, according to which all facts are supposed to be human constructions. Ontological constructivists "... confuse reality with our representation of it: the explored with the explorer, the known with the knower, the territory with its maps, America with Vespucci, facts with data, objective patterns with law statements." (Bunge 1996, 336)

48 Translator's note: Asterix is the main character of the famous French comic book series, *Asterix et Obelix.*

The drawback of Platonism "... consists of the fact that, (a) it provides no foundation for invention (as it only countenances the discovery or grasping of pre-existing entities) and (b) it postulates forms (ideas) that are divorced from material reality and are only partially accessible to experience." (Bunge 1983a, 42)

The essential drawback of Nominalism is that it confuses the signified object (the construct) with the signifying object (language, sign) and, in so doing, transforms theoretical examination into a purely haphazard manipulation of symbols.

The drawback of Empiricism is that it is not in a position to provide abstract ideas with any kind of a foundation. Among these belong, for example, several specialized branches of mathematics that are not formed through a refinement of perception, such as, for example, the so-called topological groups or mathematical spaces. (see Bunge 1983a, 42f)

Bunge provides an alternative, which he terms scientific materialism (see, for example, 1974a, 1974b, 1983a, 1996). In this, the second part of our book, we will use (and describe) those portions of his approach that are relevant for our purposes, though, admittedly, in a somewhat simplified form.

So much, then, for this excursus. Let us return, now, to the attributes of constructs.

Constructs Cannot be Observed

Since constructs are fictions, we cannot see them—they are unobservable. In simplified terms, only what emits or reflects photons of a particular frequency (somewhere between 4000 and 7000 angstroms) is observable; constructs do neither.

Constructs are not Pictures

There is an analogy between constructs and graphic pictures (as depictions of objects/events in the "Life World"). This analogy serves as the basis for propositions such as: "Science mirrors (reflects) reality", or "scientific constructs (i.e., theories) depict or portray their referents." (see Bunge 1974a, 83f)

Sentences such as these fail to get to the heart of the matter. That's because constructs are not depictions. For this reason it is useful to speak of conceptual reconstructions rather than images of reality. Constructs are constructions—artifacts—that, as a rule, must be developed with a certain degree of intellectual effort. They are not impressions or pictures that can be had effortlessly and at no cost, but rather conceptual representations of objects that cannot be observed.

The analogy between constructs and images leaves much to be desired. The reasons are as follows (see Bunge 1974a, 83f):

• Pictorial representations are themselves physical objects. Constructs, on the other hand, are "a thing of reason" (Bunge 1974a, 83).
• Images only show you what is visible. Everything else can only be hinted at, alluded to. However, the unobservable should also be grasped, after all, constructs do not stop after having skimmed the visible surface of reality: the goal is rather to represent those portions of reality that usually remain hidden

from view.

- Consequently, constructs are more symbolic (but not metaphorical), not pictorial, even when they contain certain pictorial components.
- Images, especially those made by artists, must be interpreted, usually in as many different ways as there are viewers in different moods. Constructs, however, presuppose and require that they be inter-subjectively testable.

In light of these differences between constructs and pictorial representations, it behooves us to speak of conceptual reconstructions of reality, not of images.

Animals have a picture of reality, just as people have a picture of reality. In addition, however, people also adopt what they have wrought, namely constructs as conceptual representations of those objects that cannot be observed. To be sure, these reconstructions are imperfect and, at best, only partially applicable. But, they can be tested and either improved or replaced by more appropriate representations (see Bunge 1974a, 83f).

Related are, for example, "leitmotifs [in urban planning] ..., which is to say those images that we do not have before our eyes, but rather, without being conscious of it, 'behind' our eyes. Images whose patterns we rediscover—or miss—in the layout of a city precisely because they guided our own way of seeing and the conclusions we reach. A leitmotif is ... not identical with a reproducible, viewable image..." (Schneider 1998, 124)

Only Propositions—Not Concepts—Can be "True" or "False"

Bunge makes this point in the following way: "Definitions [of concepts] are stipulations, or conventions, not assumptions. They are true by conventions, not by proof or by virtue of empirical evidence. ... In principle, nothing but practical convenience stands in the way of changing the conventional name for a thing, property, or process. In fact, such changes happen all the time. Thus in American political lingo, 'working class', state,' and 'reactionary' have been replaced by 'middle class,' 'government', and 'conservative', respectively." (Bunge 1996, 69)

And—as we have already mentioned: "Concepts ... are the units of meaning and hence the building blocks of ... discourse. We use concepts to form propositions, just as we analyze complex propositions into simpler ones and these, in turn, into concepts. *A proposition or a statement 'says' something about one or more items: it is an assertion or denial.* Even statements of possibility and of doubt are affirmations. *... Propositions are the bearers of testability and untestability, as well as of truth and falsity. That is, only propositions can be tested for truth. Concepts cannot be tested, because they neither assert nor deny anything. Hence there are no true or false concepts:* concepts can only be exact or fuzzy, applicable or inapplicable, fruitful or barren." (Bunge 1996, 49; emphasis not in original)

Thus, only propositions can be "true" or "false", because they assert or negate something. Concepts cannot be tested for their truth or falsity, because they neither assert nor deny anything. Concepts are therefore neither "true" nor "false", but rather precise or vague, applicable or inapplicable, fruitful or barren.

3.10 Consequences for Planning

Having described various attributes of constructs, in what follows we shall outline several consequences for planning; in this regard see especially also Section 3.13 (note that the fundamental connections are not only meaningful for planners, but many other professions as well):

Constructs Must be Expressed Through Language/Signs

Since constructs are unobservable thoughts it is impossible to apprehend them simply by trying to look in their direction. There is only one way to approach a construct: Constructs must be expressed in language (spoken or written) or through signs in order to be intelligible or understandable to a third party (this holds at least for so long as the techniques of mind-reading, or telepathy have not been sufficiently researched and have not yet matured [see Paulus 1994, 112]). A construct that has not been described does not exist except in the "Cognitive Apparatus" of whoever is thinking it. Put differently: A text in, for example, urban planning may contain any number of words and sentences. So long as the constructs have not been adequately described, though, this text remains, in a sense, without substance.

The Content as Well as The Range of Definitions of a Construct Must be Treated and Tested in Relation to The Formulation of a Planning Question

In a certain sense, concepts are arbitrary: concepts, like all other constructs, are unobservable cognitive fictions invented by us humans. They are—as we have seen—neither "true" nor "false". This means that even in planning, we can assert or demand the existence of any given construct. So long as one does not thereby introduce any internal contradictions, nobody can refute this effort; at best, the assertion of some construct's existence can be ignored because the concept is seen as uninteresting. Just because concepts can be invented "willy-nilly" does not mean that for planners they are, or should be, haphazard. One does not assert the existence of a useless concept. In the present context, concepts should serve mainly to aid us in solving planning tasks. This means that concepts in planning must be developed and tested in agreement and interplay with the formulation of planning questions. It is not rare that in so doing we end up modifying both.

Constructs are never all-embracing: every time we use (whether consciously or unconsciously) a construct (concept, proposition, context, etc.), we produce a certain blindness. Our perspective is limited to what we subsume in this construct. The fact that constructs are human fictions has as a consequence that, among other things, they cannot be defined "correctly" and "conclusively", in the strict sense of those words. Our definitions cannot integrate *all* theoretically possible aspects, i.e. represent *every* facet of the real world, without contradiction. After all, there are endless attributes with whose help, e.g., concepts can be defined. Constructs only partially "intersect" reality. What is included in a definition is only a subset of everything that could be included. With constructs we always produce both: the possibility for comprehension as well as blindness. Constructs therefore determine what objects and events are included in our observations during planning. They specify what is to be counted as "fact" and, in so doing, which arguments will be accepted as "penetrating" and relevant.

Working within the domain that has been defined by a construct often blinds us to the connections that exist beyond the limits of its definition. At the same time, it makes previously unknown perspectives of this domain accessible to us. These unexplored aspects uncover the possibility of new configurations, and this process continues to repeat itself along an in principle endless spiral. However, this is not a weakness in our thinking that we ought to avoid—on the contrary, this limitation is both inevitable and necessary. Without the abstraction made possible by this blindness, conceptual reflection is impossible. Our acceptance, whether it be conscious or unconscious, of the boundaries imposed by a certain construct is always only provisional. That's because it is always possible to reject a chosen description, to re-structure it and thereby to move beyond any given vantage point (see Winograd and Flores 1989).

Constructs, because they are always limited and never all-embracing, are provisional and open to modification. They always depend on a context. Major upheavals, especially in planning, arise from the modification of a construct. Examples: The (European idea of) automobile-friendly cities of twenty years ago, and the increased demand for energy today, etc. The construct "railway station", for example, has also changed. Today, we find exhibits, supermarkets, and large-format television monitors there. Similarly, we live in very different dwellings than we did fifty years ago: whether it is the size of an apartment or that we routinely plan for the inclusion of rooms for children nowadays.

Constructs Determine our Actions in Planning

Because only constructs are bearers of our knowledge, they determine and guide our actions in planning, whether we know it or not. But how could it be otherwise? After all, in planning we cannot use a "meaning" that is inaccessible to our "Cognitive Apparatus" at the moment in question. This is all to say that, for the most part, cognitive fictions, and not reality, form the basis for our actions in planning.

Of course, constructs are modified in their interplay with objective reality, the so-called facts. Constructs are, however, also modified on the basis of other, or newer, constructs—that is, without relation to the reality outside our "Cognitive Apparatus" .It follows from this that reality is not nearly as influential in planning as many lay people might think.

Some examples of constructs that guide our actions in planning:

• Whoever subsumes only objects that are incinerated somewhere or stored in landfills under the construct "waste", will—as was the case until the end of the 1970s—fail to realize the possibility of reducing and recycling waste.
• Whoever defines the constructs "urban planning", "urban construction", and "architecture", such that they only concern the distribution of space in land-use planning, construction planning, etc., as well as planning the construction of buildings or groups of buildings with their external facilities, will fail to

realize all the measures in planning that *seek to influence or regulate the behavior of those who use* these spaces (e.g., the streets, buildings, groups of buildings, external facilities, etc.). Examples of such measures include reserved street parking, car sharing, car pooling, tolls, logistical systems (management of freight traffic), etc.

- Whoever understands the construct "individual traffic" to mean only motorized individual traffic will hardly think about the needs of pedestrians when planning. Indeed, pedestrians were neglected in just such a way for many years. These mistakes are hardly marginal, as pedestrians make up fully twenty per cent of all commuters. Similarly, cyclists have played a subordinate role in traffic planning until the beginning of the 1980s. Furthermore, whoever understands the construct "individual traffic" only to refer to people commuting to their workplace will fail to think about leisure traffic. This despite the fact that in many places up to thirty per cent of all traffic serves this purpose. In the same way, for many years the construct "individual traffic" did not include freight trucks. In many places, such as Munich, this type of traffic makes up about thirty per cent of the total though.

- Whoever determines the location of hospitals based on the construct of "Central Place Theory", will—because the construct demands as much— take the number of residents who must be accommodated in a given area into account above all else. He will hardly factor the specialization that has emerged among hospitals in the past several years into his decision. The result: in Germany we are still trying to correct the mistakes produced by this kind of narrow perspective, such as, for example, with the so-called regional hospitals, of which many were planned despite the dwindling need for them.

- Whoever understands the construct "conservation" as something that only takes place in "national parks", "protected habitats", and the like—meaning, in land set aside for the expressed purpose of conservation—will fail to see the possibilities for nature conservation in daily life: saving electricity at home and at work, saving water and implementing technology that minimizes the amount of sewage generated, etc.

- Whoever favors the *model of regeneration* in the construct "sustainability"— a model in which the destruction of renewable resources cannot exceed the regeneration of those resources—will only take as much wood out of a forest as can be re-grown in that amount of time. A supporter of the *substitution model* of sustainability, on the other hand, reaches other recommendations: according to this model, substitution is permissible so long as the given goal can continue to be met; an example: if the goal of supplying us humans with energy can be met by new sources of energy, then we may deplete existing sources of oil and natural gas. Someone who subscribes to the idea of *sustaining a system* will act differently still. In this model, sustainability has been reached when the given system—"human" or "corn plant"—is preserved. For this reason, he will be satisfied if, e.g., the system is made capable of resisting contaminants from the outside.

We fashion our world, then, with the help of fictive constructs—"pedestrian zone", "region", "nature"—via language, which is to say signs. At the beginning of a planning task we must therefore clarify the central constructs before we do anything else—an essential step that is often overlooked.

There is no workable alternative to the conscious and controlled development of constructs: sharper thinking gradually leads to better constructs. A given construct is usually "better" if—as has already been pointed out—it (1) does not suffer from internal contradictions, (2) is logically compatible with as many other constructs as possible (which is to say that it manages to integrate as many [partial] constructs as possible), (3) agrees with all the available factual knowledge, and (4) is furthermore helpful in relation to the planning problem at hand.

Three Categories of Constructs: Defined, Definable, and Self-Explanatory Constructs

Working with constructs is comparably uncomplicated whenever constructs suitable to the question at hand are already available.

Accordingly, we can make a distinction between three categories of constructs:

- Defined constructs; in this case, we already have a definition for the constructs that, having already been applied successfully, as a rule goes unchallenged. Examples in construction planning include "ring armature" or "heat transmission coefficient" and in urban planning "density factor [Grundflächenzahl]" or "floor space index [Geschossflächenzahl]."
- Definable constructs; these are the constructs whose content is not sufficiently precise. In planning or designing there are ultimately always several constructs that have not been sufficiently defined. Often, those portions of the planning problem made up of previously defined constructs are few and far between. "Urban development" is one of these empty phrases. The meaning cannot be discerned without further information. Solid components are, as a rule, the city as a physical entity and a social construct. Everything else is often vague.
  Urban development can manifest itself as a result of an increase in tourism or industrial production, an increase in the amount of land set aside for recreational use, a more sophisticated public transit system, linking to a freeway, or something similar.
- Self-explanatory constructs; these are constructs for which everyone understands what is meant by them, such as, e.g., "living room", or "pedestrian zone".

A note of caution: self-explanatory constructs often turn out to be deceptive and turn out to be constructs that must be defined. When someone says "dining room", we tend to think of something in which there is a table around which are groups of chairs or other seats. However, in an, e.g., Pakistani household this is not the case.

There, there is no centralized table and the seats, if they are even present, are lined up against the wall. When we say "house", then in Germany or in Switzerland we mean a different construct than in the Philippines. Similarly, the word "planning" does not mean the same thing in Great Britain as it does in Germany, because in England there are hardly any lawfully binding plans.

Thus, even self-explanatory constructs can lead to problems, because whatever aspects of a thing are "self-explanatory" are generally limited to a certain culture—they are context-dependent. If the context is changed, then the constructs must be tested and, if necessary, changed too. Especially in international development and technical cooperation, we can often trace mistakes in planning back to the fact that this problem did not receive adequate consideration (see, for example, Hagen 1988).

3.11   Possible Mistakes

The distinctions arrived at with help of the semiotic triangle provided us with an opportunity to point out some typical sources of error that come up when dealing with constructs. In planning, these errors sometimes obviously also manifest themselves in combination with one another.

Imprecise Constructs

Constructs are often vague or hazy. Especially when they appear for the first time, they are almost always insufficiently worked out. After all: "A good idea, even if somewhat fuzzy, is preferable to an exact but pointless … one." (Bunge 1996, 61) At first, constructs are more like an unordered set of loosely connected propositions that contain more or less unclear concepts. Such "embryos" develop, if at all, by way of a combination and/or sorting out of partial components, elucidation with the aid of examples, generalizations, refinement of concepts, and testing with the aid of empirical data. But most construct-"embryos" do not mature, either because they contradict empirical data, or because the people who must deal with them do not know how to develop and refine them, or because the problems to which they relate are deemed uninteresting (see Bunge 1996, 114f). "Ironically, both intellectual immaturity and intellectual decadence share the one trait: namely conceptual imprecision." (Bunge 1996, 57) As a rule, "Avoid any words that fail to convey clear ideas: obscurity is not the mark of profundity but of confusion or even of intellectual swindle. As for fuzzy ideas—all ideas are fuzzy when newly born—try and refine them." (Bunge 1977, 8)

Sometimes, a linguistic expression is given as a name for something, but the content of the construct is not further described. When we find a word or a name for something, we at times believe that we have thus grasped the meaning of what that word designates. This phenomenon is, in part, the basis for those fashionable slogans that replace each other in planning: "Ecological" planning follows from "environmentally friendly" planning and "sustainable" urban development follows from "cautious" development. However, "Being a conceptual defect, vagueness can be reduced only by conceptual means: purely linguistic tricks will not help." (Bunge

1983b, 183) In these cases we are dealing with "embryonic constructs" that have yet to be formulated more fully.

The fashionable concept "sustainable", for example, is just such a linguistic expression; even today, if it is defined at all, it is only defined vaguely in many of those publications in which it plays a central role. Nor is it rare that a definition of the concepts being used is forgone altogether.[49],[50]

The question thus presents itself: "Are we dealing now with the role of rhetoric in planning, which one needn't take too seriously? Even urban planners step onto their soap-boxes. Successful soap-box speeches mark themselves out by constructing both speaker and audience in such a way that both parties, without ever having to say so, know not to take the other too seriously." (Jessen 1996, 17)

Most planners are highly trained at producing iconic depictions of buildings or urban spaces that rest on a factual similarity with the objects (houses, etc.) to be fabricated at a later date, i.e., at drawing pictorial plans and constructing viewable models from cardboard, wires, etc. Naturally, they use countless constructs—via different kinds of language and signs—when they do so.

What they are much less practiced at, however, is the controlled interplay of designation and semiotic interpretation (i.e., the interpretation of a construct's conceptual content). The latter serves the purpose of defining a construct in such a way that it (1) has a meaning, (2) does not suffer from internal contradictions, (3) is logically compatible with as many other constructs as possible (which is to say that it manages to integrate as many [partial] constructs as possible), (4) agrees with the available factual knowledge, and (5) is furthermore helpful in relation to the planning problem at hand. Instead, planners occasionally produce jargon—"incomprehensible murmuring" (Bußmann 1990, 361)—with the consequence that one can often only ask for a clarification of the constructs being used at the risk of ending up with nothing but a vague, nebulous mess. Since these are the constructs that help determine our actions in planning, they bring with them considerable uncertainty in terms of what the results of our planning will be.

Planners, as well as designers, must confront a second difficulty here: when developing three dimensional designs for, e.g., buildings (the classic blueprints in architecture or urban construction), then the uncontrolled use of words, i.e., free association, is an important heuristic creativity technique (see DeBono 1972). Whoever insists on inspecting every single thing or word under a magnifying glass will hardly produce any original, workable design ideas. For this reason, it is essential to decide what one is trying to accomplish: if we are trying to come up

---

49 For example, the July 1999 issue of the journal *PlanerIn*, dedicated to the Third European Planners Biennial (from 14–19 September in 1999), entitled "Sustainable Development—Challenge to the Development of European Regions", contains seventeen entries. Concepts such as "sustainable", "sustainability", etc., come up in fourteen of these articles. However, they are only defined more or less comprehensibly in two of these papers.

50 An objection might be made at this point that in many cases "sustainability" is not even at issue. Rather, "sustainability" is only a label being used to, e.g., make it easier to obtain subsidies to work on a different topic. If this is the case, it still changes nothing about the core of the problem at hand: the task of developing precise constructs has only been shifted to these "different topics".

with a design idea, then the use of undefined words is not only permissible, but also necessary. If, however, we are trying to work out a construct, then the same kind of behavior is detrimental.

Constructs—this much should have become clear by now—must be developed consciously and deliberately. In particular, it is not a good idea to leave the development of constructs up to human intuition, the part of our intellect that works unreflectively. In this case, "intuitive" must be equated with "prone to mistakes". This is because cognitive science provides us with a great deal of empirical evidence that in intuitive thinking (i.e., in thinking that is not consciously controlled) our "Cognitive Apparatus" uses innate cognitive tendencies the results of which are more a consequence of the way our "Cognitive Apparatus" works than the topic or object that we are conceptually developing—this point has already been made (see above, Chapter 2 or Schönwandt 1986). Examples include: visible images are overrated and abstract ones underrated in the formation of constructs. We have a tendency to prefer simplified constructs (patterns or connections) on the basis of "Gestalt"-recognition and the figure-ground-principle. We are often unable to appraise non-linear conceptual connections correctly, etc.

This plea for the development of constructs does not intend to deny that there indeed are situations in which the use of unclear constructs has its advantages. If unclear or ambiguous constructs are used in, e.g., the development of a design submitted to a competition in urban construction, then in some cases this ensures sufficient leeway for future uses, exactly because no precise specification was reached. In addition, judges of the contest thereby have the possibility of "reading" their own ideas "into" the given design.

One drawback remains, however: in the case of unclear constructs, it is hardly possible to discuss their meaning and content efficiently. Neither, therefore, is it really possible to improve these constructs, should they need improvement.

Ambiguous Constructs

Constructs are often ambiguous. When this is the case, a linguistic expression designates more than just one construct (concept, proposition, etc.).

The construct "job" is defined differently in a population census than it is in an employment census (see UVF 1984, 97ff). They are often mixed up, for example, when the calculated numbers are compared to each other without further elucidation. In employment censuses, as opposed to a population census, people who work in more than one place are counted more than once. This despite the fact that the equation "1 person = 1 workplace = 1 job" (Signer 1994, 61) no longer holds. This has serious consequences and effects, e.g., the space required by a firm or the volume of traffic.

The concept "workplace" comes up in at least three different technical languages:

- as office equipment (desk, chair, computer, telephone, etc.),
- as a space on a staff appointment scheme with all costs including costs aside from those associated with salaries, and

- the description of actions by a working person, although it could also be two or more working people that share a work space—a "job".

This description of the construct "workplace" also makes clear that planners do not create workplaces when, e.g., they (only) assign physical space in the form of industrial and commercial zones.

The fact that the transition from unclear to ambiguous constructs is, among other things, smooth can be illustrated with the example of mission statements in urban planning: "The spectrum of what operates as a 'mission statement' ranges from a synonymous usage for goals, principles and concepts of urban construction, urban planning and the ordering of space, to the simple labeling of trends that perpetuate themselves irrespective of our actions and the formulation of pathetic mission statements with a missionary content, all the way to a motive offer for image management and public relations strategies." (Becker et al. 1998a, 13) At the same time, appraisals of how useful mission statements are vary, on the one extreme, from: "today, developmental mission statements in urban construction are—as they have always been—indispensable preconditions for, on, or with a city" (Zahn 1998, 186) all the way, on the other extreme, to: "I would like to argue that the fiasco in urban development started at precisely the same time that we began to operate with mission statements" (Kollhoff 1998, 92).

The situation also becomes tricky when different disciplines use the same word but without noticing that they are using it to designate different constructs (see Section 3.7). Thus, the usage of, e.g., the word "space" often leads to misunderstandings. Whereas planners usually mean actual concrete space sociologist and psychologists usually mean "social space", such as, e.g., Sommer's (1969) construct of "personal space". Political scientists, on the other hand, usually mean "political space". Clearly, these are all completely different constructs.

Krätke, for example, allows us to make a distinction between four different concepts of space in planning, economics, and the social sciences:

1. "Non-Spatial" Concepts

These allow us to "tie" society and economy together in abstract space, where there are no spatial obstacles or asymmetries.

2. The Concept of Repository Space

Here, e.g., the national economic realm is viewed as a "repository" of regions, or the city as a "repository" of a number of industrial parks. This understanding of space is widespread in, e.g., traditional concepts of urban development: the city must be "filled up" with industrial parks (made up of growing sectors in the economy), with infrastructure, or with location offers (e.g., "positive location factors") of all kinds.

3. The Concept of the Relational Organizing Space of Material Objects

A central issue here is, e.g., the arrangement of functional space, commercial space, or buildings (and their respective distances from one another) within an (urban) region. This way to conceptualize/imagine space is widespread in, among other disciplines, geography and urban planning: we want to regulate the location, allocation, and size of a functional space (which, in turn, act as repository spaces

for objects/buildings), or we want to regulate the assignment of physical design-elements of an urban space (structures of buildings, among other things).
4. The Concept of Interweaving Space

Here we examine space constituted by, e.g., communication-relations as well as physico-material transfer-relations (hence, industrial supply networks, transports, the flows of traffic, etc.), or through economic control-relations (the right to decide and manage), or financial transfer-relations (the flows of capital). This conception of/image of space is relevant for, e.g., spatial exploration in the social sciences, economic and social geography, urban economy, and the politics of urban development: We concentrate on interaction-relations and interweaving connections between economic and social activities and agents in space." (Krätke 1995, 15)

Ontologically Ill-Formed Constructs

Every construct that violates the division between constructs and material objects, we will call ontologically ill-formed. If conceptual attributes are ascribed to material objects, and physical (or chemical, biological, social, or psychological) properties ascribed to constructs, ontologically ill-formed constructs result. For example: "The natural numbers are ancient" instead of "people invented the natural numbers in prehistoric times." (see Bunge 1983a, 47) Carnap mocked the ontologically ill-formed "Caesar is a prime number." Caesar was a person, hence an object, and as such he could not have also been a construct, such as a prime number.

Further examples:

- "Nature ensures that maladapted organisms die out." "Nature" is a construct and constructs do not act, therefore they cannot "ensure" anything either.
- "Nature does not need us humans." A construct, such as "nature" has no needs.
- "The people of Berlin seem to have forgotten the fact that architecture has its own language, that it communicates, and that society articulates itself through its architectural creations." "Architecture" is a construct, and constructs have no language. Rather, they are designated through language. Likewise, "society" is a construct, and constructs are unable to articulate themselves.
- "The mobility of people and freight is becoming a material object." "Mobility" is a construct, constructs cannot become material objects, they only refer to material objects.

Pragmatic Interpretation

Sometimes the content of constructs, their internal structure, is not described. Rather, they are presented (i.e., interpreted) pragmatically. By this, we mean the following: It is possible to clarify, for example, the expression "3+2=5" by counting one's fingers. Archimedes' leverage principle can be described as follows (see Bunge 1974b, 35): "If I were to sit down on one end of a see-saw and you on the other, then you will go up"; which holds true just in case the person who has uttered this sentence is the heavier of the two. Likewise, it is possible to explain the construct of coordinated public transit as follows: "When you exit from your train, then you will see the connecting train, which will depart in a few moments on the track directly across from you."

Descriptions such as these are permissible as heuristic devices (see Bunge 1974b, 35ff). However, it would be a mistake to confuse these pragmatic interpretations with constructs. Pragmatic interpretations are usually incidental. They do not describe the structure of a construct. Rather, they only point to individuals, such as observers, and to actions, such as measurements. They rely on a widespread human tendency, namely "anthropomorphism", which allows us to describe a maximum number of things with the aid of human emotions and actions. As such, pragmatic interpretations cannot replace the treatment of constructs with the aid of semiotic interpretation.

Constructs that Contain no Information about the Factual Referents to Which they Refer

An example is the concept "development." The implementation of this concept only makes sense when it is clear what—concretely, i.e., on a factual level—is developing: "Development" is the change of a property—but which factual property? In the case of a city, for example, are we dealing with the number of its inhabitants, with its financial expenditures, with the set of pollutants in its air, with the number of employees in its administration, etc.? The concept "development" does not tell us anything *per se* (see Heidemann 1993). For example: "Specific combinations of location factors in an area ("land quality") influences a region's 'potential for development.'" (Krätke 1995, 25) The same thing goes for the concepts "buffer space" and "compensatory space" in ecological planning. Without any concrete information about what it is that will be buffered or compensated, these concepts lack a factual basis.

Constructs Must Accurately Refer to Objects and Events

Occasionally we are confronted with the following situation: Constructs are formulated and they refer to objects and events. However, when the connections are re-examined, we discover that this reference is inaccurate. Differently put: A connection asserted on the conceptual level does not survive empirical scrutiny. Hence, the following three examples (see also the examples on this topic listed under section "Rules in Planning" in Section 3.12):

Example 1: The "city of short distances"

"The mission statement of the ... [city of short distances] ... rests on the at first sight plausible assumption that dense urban structures with mixed functions and high quality open spaces and facilities produce less traffic. This is an essential ingredient of standard arguments in planning. One would like to think that this assumption can be backed up by extensive research into the relationship between the structure of a settlement and the development of traffic." (Jessen 1996, 3) However: "The opposite ... is the case." (Jessen 1996, 3) The "idea that ..., neighborhoods of mixed functionality are neighborhoods of short distances [cannot] be supported by solid empirical findings. Here, much is asserted without much being known." (Jessen 1996, 4)

Example 2: Connection between the built environment and social behavior in the home and surrounding areas

A comprehensive analysis of the empirical investigations of this topic reveals that only a minimum of the theories advocated in this area withstand empirical scrutiny. For example, it is impossible to find an empirical foundation for the following assertions: crime is especially high in underprivileged neighborhoods, communication between tenants is lower in high-rises than elsewhere, etc. (see Schönwandt 1982).

Example 3: New communications technologies lead to a reduced mobility in daily life

Among other things, it was alleged that new technologies in communications would ensure that traffic on our streets as well as in the air will, if not decrease, then at least increase at a slower rate. The first empirical investigations into this subject suggest the opposite, namely, that our everyday mobility actually increases rather than decreases as a result of new communications technologies (see Zoche 2000 or Zumkeller 2000).

These examples make it clear that we should be skeptical of connections that are merely stated. But how can such statements be tested *ad hoc*. Fortunately, there are several findings in cognitive psychology that help us assess the seriousness of stated connections. When a speaker or a writer reports some empirical connections, he can do so on the basis of two kinds of information: either on the basis of data, or only by means of constructs (concepts, propositions, etc.), i.e., without any data. If there are no data at hand in the evaluation of a connection, but only constructs are used, then the human "Cognitive Apparatus" has a tendency to posit connections of notable strength. The assumed strength of the connections often exceeds the actual empirical connections a great deal (see Jennings, Amabile, and Ross 1982). This is to say that real connections are, as a rule, overestimated when they are posited only on the basis of constructs (i.e., without any data). It is therefore always important to ask oneself whether the author's or speaker's argument is founded on data or constructs. If he does not have any data available, it is a good idea to incorporate appropriate adjustments, if the author or speaker has not already done so (see Schönwandt 1986, 50ff).

Constructs Cannot be Grasped on Examination of their Empirical Basis Alone

Some work in planning essentially only describes material objects or events, meaning buildings, outdoor spaces, objects of nature (list of plants and animals), people, etc., and the changes they undergo, without paying attention to the conceptual connections that exist between them. In addition, self-explanatory constructs are often used to describe these objects, such as "tree" or "house". Such representations of the empirical constituents of constructs are a necessary part of the foundational work. However, since many linguistic expressions in planning do not denote any objects, but rather—as we have seen—designate constructs, these representations can only work in a very limited capacity.

So much, then, for the mistakes that are possible when using constructs.

Reactions to Unclear Constructs

The topic of constructs has been worked on very little in planning so far. Exceptions include, e.g., Heidemann 1990, Schön and Rein 1994, Signer 1994, or Seni 1996, each of which have dealt with this topic from different perspectives.

However, this does not mean that the corresponding deficits in the description and explanation of planning tasks—as well as other scientific domains—have gone unnoticed, or without criticism. On the contrary, a whole series of authors have singled out the problem of unclear constructs. The following six examples ought to suffice:

Example: Bächer (architecture)
This excerpt deals with planning the German pavilion for Expo 2000 in Hannover: "At the beginning of the competition the exhibition had still not been conceptualized. The judges thought nothing of it: after all, a topic had been determined: 'Human-Nature-Technology.' The Coal Union had already done an excellent job representing this in Bruxelles, 1958. The first doubts surface, if ... the distinction between a title and a concept [read: construct] is understood. Furthermore, state the supporters, neither the content nor the pavilion is the task at hand, but rather the pavilion's future use... August Everding is brought in as artistic consultant [of the jury]. He says he feels as though he has been hired to design the set of a theatre without having been told what piece will be performed there." (Bächer 1998, 1255)

Example: Möller (sustainability)
"Is sustainability merely an empty phrase?" (Möller 1999, 12); and: "Suspicion with everything that operates under the name sustainability is ... called for." (Möller 1999, 12)

Example: Strohmeyer (urban planning)
"Three types of cities emerge from the conceptual fog of contradictory insights: the closed, 'European city', ... the open city, where it remains open to whom it must open itself ... [and] the giant cities of the third world ..." (Strohmeyer 1999, 60)

The following examples show that the problem of unclear constructs is not only criticized in spatial planning:

Example: Malik (economics):
"Perhaps the most egregious lack of quality in planning strategies results from the careless use of catch-phrases and empty slogans. Again and again, in the literature as well as in the planning documents from actual practice, we find formulations that tell us absolutely nothing. As such, for example ... the foundational strategies of an enterprise are often described using formulations such as: "expanding our market-

share", "increasing our output", "forward-thinking", "holding in position", and so on. Such rewordings suffice, at most, in exceptional circumstances; they are not equipped to supply the enterprise with what, in a strategic sense, it really needs, namely a strong sense of direction... The art and necessity of the matter lie precisely in being able to say *how, with what, where, and at what cost* market shares may be increased, and then to judge if this, in light of all the factors and conditions, is really prudent, if one can afford the strategy or if it will prove to be the beginning of the end... An empty slogan cannot be concretized without further information, because it is compatible with every possible concretization." (Malik 1999, 130ff).

Example: Saltzwedel (the humanities)

"It is easy to notice when words were not born out of an effort to understand, but rather to satisfy the needs and laws of an entertainment market. And this market constantly demands the fast, clever, new... what counts is participation in the conversation, no matter what is said." (Saltzwedel 1998, 212)

Example: Sokal and Bricmont (philosophy)

"This book deals in mystification, deliberately obscure language, confused thinking, and the misuse of scientific concepts." (Sokal and Bricmont 1998, xi) "There are many different degrees of abuse. At one end, one finds extrapolations of scientific concepts, beyond their domain of validity, that are erroneous but for subtle reasons. At the other end, one finds numerous texts that are full of scientific words but entirely devoid of meaning." (Sokal and Bricmont 1998, 6) "Our goal is precisely to say that the king is naked (and the queen too)." (Sokal and Bricmont 1998, 5)

This list can be extended almost indefinitely.

3.12   Rules in Planning

In the previous chapter we discussed the semiotic triangle and, in somewhat greater detail, the topic of constructs. We also discussed several consequences that arise from these issues in planning.

However, this does not mean that it is impossible to plan if one hasn't explicitly thought about the topics mentioned above. To be sure, planning is impossible without constructs. But it does not follow that planners must be aware of this fact.

It is possible to plan even if one hasn't explicitly dealt with constructs. This can be done, for example, by calling on so-called "rules in planning". Of course, these rules are also based on constructs, but they can be implemented without recognizing the constructs that form their basis.

With this in mind, the following section will deal with such "rules in planning": What are rules in planning? What role do they play in planning? How are they formed? And most importantly, how are they founded on the stated connections that underlie them? Here, stated connections are a subset of constructs[51]: They contain (at

---

51 See Bunge 1996, 70ff.

least) two states A and B of a system and furthermore describe the "lawful" change of a system from state A to state B, etc. (for details see below, Section 3.12).

These rules always play a part in planning, and it is well-nigh impossible to plan something without using them. The same holds true for the corresponding stated connections, as they too are always involved. Of course, a given stated connection can be more or less well-founded. Moreover, rules and the stated connections that belong to them determine the methods we use in planning, and therefore our actions in planning.

The fact that rules in planning are based on stated connections does not mean that the former can, strictly speaking, be derived from the latter. The idiosyncratic nature of the intersection between rules and stated connections brings with it that the "functioning" of rules does not tell us anything about whether the stated connections that form the rule's basis are in fact accurate or not. What's more, the desired outcome can, among other things, be attained on the basis of partially inaccurate stated connections.

Rules Underlie Planning

As has already been pointed out, planning is usually a matter of improving a state of affairs that is deemed disadvantageous. In order to achieve the desired state, planners tend to suggest actions in the context of the "instructions" developed by them— actions that are often called "measures". Examples of measures include construction of a by-pass road, the designation of some area as a nature habitat or the public subsidy of monthly or yearly public transit passes.

If we bracket out all the components in these measures that are specific to an individual case, we are left with the rules in planning that form the basis of these measures; after all, planners do not design their proposed solutions to a planning task from the ground up every time. Rather, they have at their disposal professional experience (including knowledge of rules) that they call upon when working on a specific planning task.

(The following comments on the topic of rules are based in large part on Bunge 1983b, 1996 and 1998.)

What are rules in planning?[52] They are instructions that can be used in similar planning situations to decide how one might bring about a desired result. "A *rule* is an instruction for doing something definite with things, processes or ideas." (Bunge 1996, 73) "A rule prescribes a course of action: it indicates how one should proceed in order to achieve a predetermined goal. More explicitly: a rule is an instruction to perform a finite number of acts ... with a given aim. ... In contrast to law [statements], which say what the shape of possible events is, rules are norms." (Bunge 1998, 147) Rules are therefore instructions for what to do, be it intellectual, manual, or in the form of communication (i.e., interaction) with others. If we are dealing with conceptual problems, they guide our thinking. If the problems are practical in their nature, then they determine our actions. As opposed to stated connections, for

---

52 A "rule" is the concept usually used by science in this context; see, for example, Stone 1988, 231ff. or Bunge 1996, 73ff.

example, rules describe nothing, they explain nothing, and they predict nothing: they instruct.

Examples of such rules include: "In order to protect the extant flora and fauna, set certain sections of the environment aside for protection", "to convince more motorists to switch to public transit, subsidize monthly or yearly passes", or "in order to reduce the through-traffic in a region, build a by-pass road". At their core, rules therefore include at least two states, (a) an unwanted and (b) a desired one, (in the examples just mentioned, this would be "a great deal of through-traffic" and "less through-traffic"), as well as (c) the actions that people can take to transform the first state into the second, in this example by building a by-pass road.

These rules have the following abstract structure: In order to reach the desired state S, use method M. Or, in order to avoid S, don't do M.[53] In the example above, "subsidize the monthly and yearly passes in order to convince more motorists to switch to public transit", the subsidized passes would be the methods and the motorists switching from their cars to some form of public transit would be the desired goal.

Rules are Always Used When a Situation is to be Purposefully Altered

Rules are not only used in spatial planning (architecture, urban construction, urban planning, landscape planning, regional planning, etc.). Rather, they are always used when we are dealing with a situation in which something needs to be purposefully altered, a problem needs to be solved, a team needs to be led, etc. This means that they always come into play when measures are used to bring about something; when a technical, social, or some other system needs to be controlled or changed, or when a new product is to be created. As such, rules exist in several domains, such as:

• In the construction of buildings: "In order to avoid damage to walls that are

---

53 At the basis of this section lies the so-called ends-means-approach, which can be formulated as follows: a desired end (goal), which will not come about on its own, can only be reached through our actions if we implement certain means.

When using the ends-means-approach, a series of problems emerge that, nonetheless, do not put the core of this approach into question: ends and means are not simply given, nor are they value-neutral. To agree on a certain end with those involved and affected by planning is one of the most difficult aspects of planning. Ends and means change; after all, the knowledge-base, perspective, and preference of those involved often change as time goes by. Moreover, ends can consist of several sub-ends, which often contradict one another. Furthermore, several means are often possible, each one of which competes with the rest. Additionally, the relationship between the desired end and the planned means can be simple, such as if a direct connection between both is assumed; for example: "In order to decrease the number of pollutants that are produced by motorized traffic on a street (nitric oxide, hydrocarbons, etc.) (=end), reduce the speed limit prescribed for this street (=mean)." Or it can consist of a number of connections that together form a complex system. And, especially in processes that form a feedback loop, it can be difficult to determine what is an "end" and what is a "means", because a given end may also be the means with which to achieve some further ends (see for example Forester 1993, 20ff, Alexander 1992, 54ff, Banfield 1973, 139ff, or Schönwandt 1999).

made of material whose integrity declines over time, build them such that the connections between them are as short as possible and/or they have a sufficient number of expansion joints."

- In the treatment of patients: "Use antibiotics to heal bacterial infections."
- In economics: "Encourage demand to stimulate economic growth."etc.

In short: Everyone who wants to change something in the real world by way of planning, makes use of some rules of planning (more generally: rules of technology or social technology) that he thinks will suffice to bring about the desired state of affairs.

Rules are Based on Stated Connections

Rules are based on "underlying" stated connections. An example: the rule is: "in order to increase the number of people who use public transport (=goal), reduce the hassle of making a transfer: reduce the waiting time, create comfortable waiting areas in bus stops and train stations (=method)." A stated connection that reinforces this rule would, for example, be the following: "public transit routes that require passengers to make several transfers are used less frequently compared with direct routes that require no transfers; the reason: it is often necessary to wait a long time for a transfer, and the waiting areas are often uncomfortable."

Of course, the veracity of the stated connections that underlie the given rules vary rather widely. Sometimes they are merely suspicions that a planner needn't even be conscious of. At other times they are scientifically justified. Scientifically justified stated connections are constructs and, as such—to put it generally—empirically tested hypotheses that, if necessary, are embedded in a theory and represent some regularity. As a scientific statement (i.e., construct), they tend to have the following structure: In a system of kind S, the state or event B follows—either always or with some specific frequency—from the state or event A. Moreover, it does so consistently with a regularity that varies with respect to time. They thus contain (at least) two states, A and B. Moreover, they describe a temporally regularly occurring "lawful" change from the first state, A, to the second state, B, etc. Further ways of saying the same thing include: "A causes B, B follows A, or A makes B likely" (more specifically: "The likelihood of B, given A, is significantly higher than zero") (see Bunge 1996, 73). Aside from the description of a process as a succession of different states of a system, stated connections also include portions of the relations described in Section 3.3 above, especially statistical, functional, probabilistic, and causal relations. Note also that the last two relations are both called mechanisms (see Section 3.4).

As such, stated connections are a subset of all propositions. Namely, they are ones in which concepts are combined to form propositions mainly using statistical, functional, probabilistic, or causal relations.

The Difference Between Rules and Stated Connections

The structural difference between rules and stated connections is as follows: A stated connection is—as described—the representation of a system and describes the "lawful" change of this system from a state A into a different state B (as well as, in some cases, the further state C, etc.).

Rules, on the other hand, create a connection between certain different states of the given system in question (such as, for example, a state of affairs deemed disadvantageous as well as a desired state, the goal) on the one hand, and a *set of possible actions*, which can be carried out by agents to change the system and bring about the desired state on the other hand.

Therefore, whereas a stated connection ascertains that (and at times also how) a system changes from state A to state B, rules not only concern the states of a system, but also incorporate the actions that agents can carry out in order to change the system from state A into state B. This means that every rule is at base the description of the change in a system from state A into state B and therefore contains a stated connection—even if the latter is not formulated explicitly.

For this reason it is always possible to discover or expose the stated connections employed in a given planning task: this can by accomplished by

1. finding out what the initial conditions are, as well as
2. which goals ought to be reached using the measure, and
3. analyzing precisely what planning measures are being proposed for implementation.

If, for example, we build a shopping center (=planning measure) in what is understood to be an underprivileged neighborhood (=initial condition), in order to gentrify this neighborhood (=desired state, goal), then we are using the following (simple) rule: "In order to gentrify a neighborhood, build an urban or shopping center there." This rule is based—either consciously or unconsciously—on the following (simple) stated connection: "Newly built urban or shopping centers lead to changes that gentrify the neighborhood in question."

Towards the Construction of Rules

How are rules in planning constructed on the basis of stated connections? They are formed by deploying a rule for at least one of the theses contained in the stated connections. Table 3.3 shows some examples (see Bunge 1983b, 371).

The following example shows how a rule can be formed from a stated connection: The foundation is the previously cited thesis: "New York is currently [experimenting] with large signs in the streets that measure air pollution and that (as has already been successfully tested in Sao Palo) will move commuters to switch to trains or buses." (Mönninger 1999, 19) It is possible, for example, to extract the following mechanism (mechanisms are a subset of all stated connections) from this formulation (although we will not do so in the detailed form of A leads to B, B leads to C, C leads to D, etc.): "The displayed air pollution figures cause commuters to switch to trains and

Table 3.3   How are rules in planning formed on the basis of stated connections?

Stated Connection		Rule in Planning
When A happens, B transpires.		In order to achieve B, do A.
When A rises, B falls.		In order to lower B, raise A.
The degree of change in A corresponds to the change in B.		In order to change A, regulate B.

buses." This mechanism may form the basis for the following rule: "In order to move commuters to switch to trains or buses, post large signs along the streets that display the amount of air pollution."

Rules Cannot be Stringently Derived from Stated Connections

When we say that a rule in planning is based on a stated connection we do not mean to imply that they can be stringently—logically, deductively, etc.—derived from those statements. This is to say that there is no one-to-one-correspondence between stated connections and rules in planning.

There are three primary reasons for the absence of such a one-to-one-correspondence:

The first reason is that there are many stated connections that do not form a basis from which it is possible to formulate a rule in planning. As such, new insights into, e.g., the physical or chemical properties of the materials (metals, synthetic materials, etc.) used by vehicles in personal traffic as well as the effects connected to them can—as has already been stated above—rarely be directly reformulated as rules for planning that have the capacity to improve the lives of citizens. Such stated connections usually have too "fine-grained" a resolution.

A second reason is the following: If it is possible to show, for example, that a rule is based on a certain stated connection, then this stated connection usually satisfies the rule only in part. Other aspects are usually needed as well: fundamentally, an assertion such as "if method M, then desired state S" can mean that it is possible to achieve S by using or implementing M. But even if this stated connection were accurate, it would not tell us anything about whether or not *it is even possible to bring about M*. Two examples of this are: "If the conditions of all locations were everywhere identical for a firm, then businesses would never relocate jobs because there would no longer by any reason to do so". Or: "If every means of disposing plutonium were sure to remain secure for some seven hundred thousand years, then this dangerous substance would not cause any harm". In both cases it would hardly be possible to bring about the given situation: in all likelihood there will never be "identical conditions of locations the world over", nor will there ever be "ways to dispose of plutonium that remain secure for some seven hundred thousand years".

Rules cannot be stringently derived from stated connections for a third reason: every stated connection is, either actually or potentially, the basis not only for one, but for two different rules in planning. One, which prescribes what must be done to achieve a desired state of affairs, as well as its "dual rule", which describes how one ought to act if one wants to avoid the corresponding outcome (in this case we are dealing with two different goals).

If we look at a simple stated connection, namely "when M, then S", then this single assertion forms the basis for the following pair of rules:

Rule 1: To achieve S, do M, or ensure that M transpires.

Rule 2: In order to prevent S from happening, avoid doing M, or prevent M from happening.

An example from the domain of economics includes the well-confirmed stated connection: "High unemployment leads to frozen or reduced salaries. Low unemployment leads to higher salaries." Differently put: "When unemployment is on the rise, salaries are not. If unemployment falls, then salaries rise." This assertion is suggestive of two rules:

Rule 1: In order to keep salaries low, raise unemployment.

Rule 2: In order to raise salaries, create new employment opportunities.

The stated connection does not help us decide which of these two rules we should use. The given goal—"salaries should remain low" or "salaries should rise"—must be fixed by way of further deliberation over, e.g., whose interests should be given priority.

As far as the construction of the "dual rule" goes, a further restriction must be kept in mind as well. Often, the same stated connection which serves as the model for the rule "in order to achieve S, do M" also supports—as has been described above—the corresponding dual rule: "in order to prevent S from happening, avoid doing M." However, in some cases the underlying stated connection does not provide a basis for this negative assertion. That is because the "not-S" does not just mean "anti-S", but it can also mean the absence of S. An example: Let us assume the assertion, "the output of an economic system can be described as a *rising* function of capital and labor." Two rules can be formed on the basis of this assertion: "In order to increase the output of an economic system, increase the capital or the labor or both", and "in order to decrease the output of an economic system, decrease the capital or the labor or both." However, the question of what leads to a *decreasing* output of an economic system cannot be answered based on the above stated connection alone (see above). It only mentions a "rising function", there is no talk of any "falling function." Hence, the latter of the two rules above is not covered by the stated assertion and therefore cannot be formulated legitimately.

The Moral Ambivalence of Rules

The fact that a stated connection can serve as the basis for two rules lies at the root of the moral ambivalence of planning. This is in contrast to the moral univalence of basic research. The same thing holds for all the technological or social-technological rules; technology and social technology are correspondingly ambivalent. A stated connection describes, explains, or predicts. As such, it is morally[54] neutral (which is not to be confused with value neutrality[55]). A rule in planning, on the other hand, is a prescription for action. This rule can prescribe morally "right" or "wrong" actions. If the desired state S and the method M that is used to achieve S is morally "right", then the stated connection "When M, then S" serves as the basis for the morally "right" rule: "In order to achieve the desired state S, do M." If, however, the desired state or the method are morally dubious, then the first rule is morally wrong and we should use its dual rule instead. For example, assume that the stated connection is as follows: "Construction firms that belong to a cartel usually enjoy higher profits than firms that are not members of a cartel." It would be morally (and legally) wrong to act according to the rule for which this stated connection serves as a basis: "In order to enjoy higher profits for your construction firm, join a cartel." The application of a rule can therefore be morally reprehensible as well as morally justified.

While stated connections are therefore morally neutral, rules are not. This is because goals and methods, while they might benefit some people economically, morally, or otherwise, can also harm other people. It follows from this that we must develop a moral justification for every application of a rule. Goals and methods must be examined in terms of their moral (i.e., ethical[56]) advantages and disadvantages and the given preferences must be disclosed.

Rules and Stated Connections Must be Disclosed and Tested

An essential reason why rules and stated connections must be disclosed and tested is that, in the long run, rules for planning can only "function" properly when the stated connection that form their basis are at least in part accurate. For this reason, the conscious formulation and testing of rules for planning and the stated connections that form their basis must be part and parcel of the standard for professional work in planning.

A further reason for this testing is as follows: Learning rules for planning is an important part of the professional training and activity, but it is also relatively simple because, in order to use a rule, it suffices that one *knows how to implement it*. What is not absolutely necessary is to know, e.g., why or even how the rules function and what *they individually bring about* (see Bunge 1983b, 253). The implementation of a rule also does not require a careful analysis of the planning task, i.e., the initial conditions.[57] This brings with it the danger that rules are implemented unreflectively, meaning that there is a danger of not checking whether the rule that is used when working on a planning task is even appropriate to the particular case in question.

---

54 Moral questions are questions about how one ought to behave when values and interests are in conflict; e.g. when one's self-interests conflict with the interests of one's fellows. As such, they are markers of social situations.

55 In planning, we cannot avoid working with simplifications and reductions. Moreover, our limited perception only gives us access to a "filtered" version of the world. All of this implies the inclusion of values, which means that planning cannot be value neutral.

56 Ethics is the science of morals.

57 Empirical studies show that in most situations the initial conditions are not sufficiently analyzed in planning (see, for example, von der Weth 1999, 456).

Naturally, in actual practice it sometimes happens that a planner plans with the aid of inappropriate rules, without this leading to noticeable mistakes in planning or the planner being criticized. An important reason for this is that only a few solutions of planning are systematically evaluated and compared to the original intentions. For this reason, possible mistakes in planning that stem from the implementation of inappropriate rules often go unnoticed and therefore cannot be avoided the next time.

There is also a second reason why a conscious reflection on the rules and stated connections should take place, namely in relation to the testing of concrete actions in planning. This is because it is by no means always the case that a given action in planning actually corresponds to the verbal instructions that were given for this action. In the social sciences (see, for example, Deutscher 1973 or Nisbett and DeCamp Wilson 1977) it is well known that for most people there is a difference between their actions and their verbal descriptions of those actions. Argyris and Schön (1978) have, through their empirical studies of organizations, shown that this is also a problem in the implementation of rules. For this reason, not only the rules and stated connections that a planner *claims to form the foundations* of a planning task should be examined and tested. Rather, the degree to which the planner's verbal explanation corresponds to the actual actions taken in planning must also be examined and tested. If these tests of the concrete actions that are taken in planning are forgone, then there is a danger that in some cases only the verbal explanations are corrected whereas little changes in practice.

Rules and stated connections should therefore be brought to the surface rather than being viewed as obvious or even irrelevant. This much is clear: Knowledge and implementation of stated connections that are well-founded increase the chances of successful action—where stated connections are "well-founded" especially then when they do not suffer from conceptual contradictions and when they conform to what is factually known. The conceptual and empirical testing of the given stated connections then allow us to improve on corresponding rules as well as to make their implementation more secure.

On Testing Rules and Stated Connections

A rule in planning can, in principle, be justified in two ways:

- practically, for being "useful" or "practical", meaning by its success[58], or
- conceptually, meaning by way of agreement with a stated connection.

The conceptual justification of a rule for planning consists of showing *why* it causes something, and this is done with the aid of a stated connection that is as well-confirmed as possible. This presupposes that the given stated connection is, on the one hand, formulated clearly and, on the other hand, as empirically well-confirmed as possible.

---

58 A rule is successful just in case the interventions it gives rise to bring about an outcome that corresponds to the planner's original intention (see Schönwandt 1999).

Clear Versus Unclear Stated Connections

A stated connection is clear when its depiction and content are linguistically precise and logico-conceptually consistent, meaning free of any internal contradictions.[59]

The problem with rules in planning that are based on vague or diffuse stated connections is that—in comparison to stated connections that have been subjected to tests—we usually know very little about the effects that their implementation bring about, whether they are desirable or undesirable.

If a stated connection is used unconsciously, or it can only be formulated in a vague and diffuse way, then we will call the corresponding rule "conceptually unsupported".

If a rule in planning is conceptually unsupported, yet still efficient, if it "functions" despite the fact that we do not know why—meaning, when we cannot formulate a stated connection to form the basis of a rule—then we shall call that rule a "rule of thumb".

Empirically Founded Versus Empirically Unfounded Stated Connections

A stated connection that forms the basis of a rule can either be supported by empirical investigation or unsupported. Admittedly, in planning we deal, among other things, with the implementation of proposed solutions that are completely novel. These solutions also contain rules that are based on stated connections. What remains missing in this case is the possibility of empirically testing the given stated connection, because there are as yet still no empirically testable cases in which the proposed solution has been implemented.

Irrespective of this, there are nonetheless many examples showing that a number of stated connections that are used in planning are not empirically supported; two examples ought to suffice (further examples for this have already been described in Section 3.11):

Example 1: Nature Conservation

To put it bluntly, nature conservation has the primary goal of preserving flora and fauna in need of conservation. To do this, certain areas are designated as conservation zones. In practice, the location of flora in need of conservation is the essential criteria used to determine where these conservation zones should be. The tacit assumption (stated connection) here is that the fauna in need of conservation are located in these same areas. However: The zoological aspect of nature conservation is often overlooked. Empirical investigations in corresponding areas in Upper Franconia show that in the majority of cases the zoological and botanical significance of an eco-system do not overlap. Crucially, habitats of highly endangered animal species

---

59 It is difficult to formulate a stated connection in such a way that it is in conclusive agreement with *every* other stated connection. Strictly speaking, this is actually impossible because there are always alternative explanations for events that take place in the real world. What matters is replacing stated connections that, upon closer inspection, reveal themselves to be inconsistent, with others that are conceptually *as consistent, i.e., free from internal contradictions, as we can reasonably expect.*

that overlap with areas in need of botanical conservation are disproportionately rare (see Reck 1994).

Example 2: Urban Sprawl and Increase in Automobile Traffic

For years now, people have been migrating from city centers out to the periphery. The current view holds that, since this tends to augment the distance between the people's place of residence, work, shopping, and recreational activities, automobile traffic increases accordingly. A study conducted in Bremen, however, discovered the following: The distance between workplace and place of residence within the city of Bremen and the surrounding villages (those within a thirty kilometer radius of the city) have not really increased between the years 1970 to 1987. "In the same time span, the opposite trend can be perceived in almost half of all the neighboring villages: the distances between work and home have decreased considerably. At the same time, those who lived and worked within the city of Bremen had to travel an average of 1.2 kilometers more to get to work in 1987 than they did in 1970. The reason: Firms historically located in the cities reduced their workforce or closed down. New positions, on the other hand, were created primarily on the periphery. In some cases, the sprawl of industry therefore brought the workplace closer to people's homes again. As such, the assertion that the looser settlement structure of the suburbs forces people to drive their cars more is no longer valid—at least not for the city of Bremen." (Wittmann 2000, 106; see especially also Bahrenberg and Albers 1998)

These examples make it clear that we should be skeptical of stated connections that are merely propounded without any empirical proof.

Rules in planning that are based on empirically well-founded stated connections we shall call "empirically supported rules".

Some Further Remarks on Rules

The Construct "Truth" Does Not Apply to Rules

About stated connections, as well as about propositions in general, we can say that they are accurate, i.e., "true" (for a definition of the concept "truth" see, for example, Bunge 1974b, 81ff or Groeben and Westmeyer 1975, 142ff). Since rules for planning are instructions and not stated connections, they cannot be either "true" or "false". Rather, rules are more or less relevant in connection to a given planning question and the possibility of reaching a desired state with their help. Moreover, rules can be efficient[60] to different degrees; where "efficiency" is made up of efficacy,[61] meaning effectiveness, paired with a minimal required effort and risk.

---

60 The formulation "rules are efficient" is short for: rules are catalysts for actions that, having been implemented, bring about outcomes that, measured against the required effort, etc., are more or less efficient.

61 Of course "efficacy" as well as "efficiency" must be analyzed separately, not in the least because they are based on axiological, i.e., ethical, principles: how effective, for example,

The Efficacy of a Rule Tells us Nothing About Whether the Stated Connection that Forms its Basis is in Fact Correct

Since the construct "truth" does not apply to rules, the practical success that a rule helped us to achieve does not give us reason to believe that the stated connection that forms the basis for this rule is in fact accurate. Similarly, failure does not give us reason to believe that the stated connection is inaccurate. Hence, the stated connection that forms the basis for a rule can be inaccurate, and yet the rule may still lead to success, just as a rule can lead to practical failure even though the corresponding stated connection is essentially accurate.

The most important reasons for the irrelevance of praxis in the validation of stated connections are the following: In real situations, the relevant characteristics are seldom adequately known and controlled with sufficient precision. In addition, planning tasks usually do not leave us with the time for detailed examinations in which individual variables are grasped one at a time and tied to a stated connection. The desire, or the need, to act as efficiently as possible often leads to the simultaneous implementation of several rules, and it is often the case that these rules compete with one another. If the results are satisfactory, then the planner who implemented several rules simultaneously cannot know for sure which of these rules was efficient. And as a corollary: If the results are unsatisfactory, how could he possibly know which rule or which stated connection that forms the basis of a rule is to blame?

In the daily activity of planners it is hardly possible to carefully distinguish and control the relevant variables and critically examine the connections between these variables—this is reserved for empirical analyses. This is also the reason why the stated connections used in planning cannot be tested by architecture and planning firms or by planning boards. Only systematic investigation can determine if a stated connection is accurate or not (see, for example, Campbell and Stanley 1966 or Patton and Sawicki 1993).

Stated Connections that Are Only Partially Accurate Can Also Serve as the Basis for Efficient Rules

Since there is no one-to-one correspondence between rules in planning and the stated connections that form their basis, a rule is sometimes based on an inaccurate stated connection, though it "functions" nonetheless; this happens for the following reasons.

First, a stated connection that is essentially false may nevertheless contain a grain of truth somewhere. If this grain of truth is the only portion of the stated connection that is used as the basis for formulating a rule in planning, then this rule may still help us to achieve a desired goal. After all, a stated connection is usually an arrangement of individual assertions and sometimes it is enough to achieve a desired state if only some of those individual assertions are accurate or at least very nearly so. It is only essential that inaccurate elements are either harmless or not used.

---

is a rule of planning that, although it does provide help in bringing about a certain desired state, has as a side effect an increase in unemployment or pollution?

A further reason for the possible practical success of an inaccurately stated connection is as follows: The accuracy requirements in practical planning are often much lower than those that are needed for research. Moreover, profound and complicated stated connections are sometimes inefficient in planning because too much work is required to achieve results using them. In some cases, increased precision is simply pointless because it is unnecessary. A well-known example here is the Apollo missions to the moon, which used primarily Newtonian mechanics rather than the more accurate relativistic theories. Precision, an important goal in research, is sometimes inappropriate in practice. In planning it is therefore important to choose that degree of precision or detail which is best suited to answering a given question at hand—meaning one that is neither too coarse not too fine-grained.

For these reasons—i.e., the implementation of only a portion of a stated connection and the often lower precision requirements that exist in practical planning—different, possibly even competing, stated connections can lead to the same outcome in actual practice. This is to say that it is not always important to use "correct" stated connections as a basis for a rule; if the useful result of two competing stated connections is identical, then the planner can implement whichever stated connection forms the basis of the rule that is best suited to working on a particular question at hand.

Summary

Rules for planning are based on stated connections, but there is no one-to-one correspondence between the two. Nonetheless, knowledge and use of well-founded stated connections increases the chances of successful actions in planning. Rules, as well as stated connections, ought therefore to be brought to the surface and tested. They should not be thought of as being self-evident or, worse, irrelevant.

3.13   Conclusion

The second part of this book demonstrated the significance that the semiotic triangle—and therefore the topic of language/signs, objects/events ("facts"), and especially constructs—has for planning.

In planning, it is particularly important to pay attention to the construction and testing of constructs: they are the bearers of our knowledge as well as the conceptual core of a planning task. More than anything else, they guide our actions in planning. As such, they offer insight and direction.

When working on a planning task, planners should therefore treat the following topics and be able to answer the corresponding questions (whose content partially overlaps) with sufficient precision. However, not every aspect is of the same relevance in every planning task; it is strongly recommended that readers discard individual aspects only after sufficient testing anyhow.

Questions for a planning problem:

- What is the planning problem? More specifically: What is the situation that has been deemed disadvantageous and that should be improved? Which goals should be reached? And/or: Which state of affairs that is deemed advantageous should be maintained?

- Do others who are involved formulate the planning problem differently, even though they are dealing with the same issues? If so, how and why?

Questions (only) for concepts:

- Which (key) concepts were used to describe the planning task? Which attributes were used to define these concepts? Were negations ("not-A") falsely drawn into the definition?
- Are the definitions of concepts conducted with an eye towards the formulation of a planning task and are they helpful for treating those planning tasks? Does, for example, the "resolution" of the defined concept fit with the question at hand?
- Are "final" or "true" definitions sought in vain?
- Is predication (the assignment of attributes) used to "define away" a planning problem?
- Are we ensuring that the initial definitions of concepts are not unknowingly/ silently changed in the course of the planning process, such as, for example, if in the assessment procedure process attributes (i.e. assessment criteria) other than those used in the initial definition of a concept are used "unawares"? (What is called an "attribute" in the definition of a concept is designated as an "assessment criteria" in the assessment process.)

Questions for constructs (concepts, propositions, contexts, theories):

- Do the constructs have a meaning? To what extent are purport (precursor constructs), intension (core construct, content), import (implication), extension, and reference explained with sufficient clarity and completeness?
- In what manner are the symbols being used (such as, for example, the linguistic expression "urban development") interpreted semiotically? Are we dealing with conceptual, factual, and/or empirical interpretations?
- Are iconic representations confused with semiotic interpretations in the formation of constructs?
- What are the relevant "facts" and empirical data in a planning problem? Is the referent of a given construct evident (are there corresponding reference and evidence classes)? Are there any empirical data that provide evidence for the propositions or theories being used?
- Constructs emphasize certain focal points and bracket other topics out: is this emphasis and bracketing tested in order to ascertain whether topics important to the treatment of the planning task are hidden?
- Which stage of construct formation is reached: the list/inventory, sketch/ diagram, specific theory/theoretical model, generic theory?
- Are imprecise, unclear, or ontologically ill-formed constructs avoided? Likewise, are pragmatic interpretations as well as constructs that contain no information about their factual referents avoided?

- Is the fact kept in mind that in planning several (at times inconsistent) constructs (some of which determine our actions whereas others are provided as verbal explanations of those actions) are often and unconsciously used at the same time?
- Are metaphors or analogies used as substitutes for constructs? Are all the possible errors that this might produce kept in mind? (Analogies are not isomorphisms, and descriptions of constructs with the aid of metaphors/analogies are more resistant to change than those that are, e.g., formed by way of predication.)
- Constructs play a role in planning on at least two levels: on the level of the given concrete formulation of a planning question as well as on an "underlying" level, as an "approach to planning" and a part of the "planning world": various approaches to planning have different conceptual and methodological commitments that limit the scope of our actions in an actual planning task. With this in mind: in the context of a given planning task, are the various possible approaches to planning scrutinized and are their advantages and disadvantages examined?
- Is the fact that people have, among other things, a tendency to commit certain mistakes (to fall into certain cognitive traps) when forming constructs kept in mind?
- Are constructs consciously defined ambiguously (which should not be confused with "imprecisely")? If so, why?
- Do others who are involved use different constructs (concepts, propositions, contexts, theories, metaphors, analogies)? If so, which ones and why?

Questions that deal especially with explanations, i.e., mechanisms and forces:

- What are the relevant mechanisms specific to the system that are operative in a planning situation?
- On the basis of what information are these mechanisms assumed to exist? On the basis of so-called dynamic descriptions, or are kinematical descriptions, statistical correlations, assignment to classes, teleological, functional, or tautological descriptions, "narrative explanations", empathetic explanations used instead?
- Which mechanisms should be used in the context of the planning measures?
- Is a combined top-down and bottom-up approach used when dealing with mechanisms that are supposed to influence the behavior of people? Is the fact that we are shaped by our environment at the same time as we shape our environment taken into account?
- Is there empirical evidence for the effectiveness of the mechanisms that are supposed to be used?
- Are forces that can alter either the speed or the manner of a mechanism's operation involved in the planning situation? If so, which ones, and what is to be done about these forces?
- Do others who are involved use different explanations (descriptions of mechanisms, i.e. forces)? If so, which ones and why?

Questions for the rules in planning:

- What are the rules in planning with the help of which the planning problem is to be treated?
- What are the stated connections that form the basis of these rules in planning?
- How well-founded are these stated connections and therefore the rules in planning of which they form the basis? Are the rules conceptually and empirically justified, or are we dealing with "rules of thumb" and therefore—as Heidemann (1985) formulated it so poignantly—with rumors and speculations?
- Do others who are involved use different rules in planning? If so, which ones and why?

Answering these questions can help structure and guide the work we do on the content of constructs. This manner of proceeding can furthermore contribute to restricting our use of those concepts that lack a discernible meaning or clarity and are therefore usually senseless.

Bibliography

Alexander, E.R. (1984), 'After Rationality, What?', *Journal of the American Planning Association* 50: 62-9.

Alexander, E.R. (1992), *Approaches to Planning; Introducing Current Planning Theories, Concepts and Issues* (Luxembourg: Gordon and Breach Science Publishers).

Alexander, E.R. (1996), 'After Rationality: Towards a Contingency Theory of Planning', in Mandelbaum, Mazza and Burchell, pp. 45-64.

Allmendinger, Ph. (2000), 'Planning in the Future: Trends, Problems and Possibilities', in Allmendinger, Ph. and Chapman, M. *Planning Beyond 2000* (Chichester: John Wiley & Sons), pp. 241-274.

Allmendinger, Ph. (2002), *Planning Theory* (New York: Palgrave).

Allmendinger, Ph. and Tewdwr-Jones, M. (eds) (2002), *Planning Futures: New Directions for Planning Theory* (London: Routledge).

Anderson, J.R. (1989), *Kognitive Psychologie* (Heidelberg: Spektrum der Wissenschaft Verlagsgesellschaft).

Argyris, C. and Schön, D.A. (1978), *Organizational Learning* (New York: Addison-Wesley).

ARL (Akademie für Raumforschung und Landesplanung, Landesarbeitsgemeinschaft Baden-Württemberg) (2000), Stellungnahme zum Landesentwicklungsplan Baden-Württemberg – Anhörungsentwurf.

Athearn, D. (1994), *Scientific Nihilism: On the Loss and Recovery of Physical Explanation* (Albany: State University of New York Press).

Bächer, M. (1998), 'EXPO 2000 – Made in Germany', in *Deutsches Architektenblatt* (10): 1255-1256.

Bäcker, A. (1996), *Rationalität als Grundproblem der strategischen Unternehmensplanung. Ein Beitrag zur Erklärung und Überwindung der Rationalitätskrise in der Planungstheorie* (Wiesbaden: Deutscher Universitätsverlag).

Bahrenberg, G. and Albers, K. (1998), 'Die Kernstadt, das Umland und die Folgen des Trends. Führt die Suburbanisierung zu mehr Autoverkehr?', in *Mitteilungen der Deutschen Forschungsgemeinschaft* (4): 4-6.

Banai, R. (1988), 'Planning Paradigms: contradictions and synthesis', in *Journal of Architecture and Planning Research* (5)1: 14-34.

Banfield E.C. (1973), 'Ends and Means in Planning', in Faludi, pp. 139-149.

Bartlett, F.C. (1932), *Remembering: A Study in Experimental and Social Psychology* (Cambridge: Cambridge University Press).

Bechtolsheim, M. von (1993), *Agentensysteme: verteiltes Problemlösen mit Expertensystemen* (Braunschweig: Vieweg).

Becker, H., Jessen, J. and Sander, R. (1998), *Ohne Leitbild? – Städtebau in Deutschland und Europa* (Stuttgart, Zürich: Karl Krämer).

Becker, H., Jessen, J. and Sander, R. (1998a), 'Auf der Suche nach Orientierung – das Wiederaufleben der Leitbildfrage im Städtebau', in Becker et al., pp. 10-17.

Bem, D.J. and Allen, A. (1974), 'On predicting some of the people some of the time: The search for cross-situational consistencies in behavior', in *Psychological Review* 81: 506-520.

Bem, D.J. and Funder, D.C. (1978), 'Predicting more of the people more of the time: Assessing the personality of situations', in *Psychological Review* (85)6: 485-501.

Bense, M. (1971), *Zeichen und Design: Semiotische Ästhetik*. Baden-Baden: Agis.

Blotevogel, H.H. (1995), 'Zentrale Orte', in Akademie für Raumforschung und Landesplanung (ed.) (1995), *Handwörterbuch der Raumordnung* (Hannover: Akademie für Raumforschung und Landesplanung), pp. 1117-1124.

Bonny, H.W. (1998), 'Funktionsmischung – zur Integration der Funktionen Wohnen und Arbeiten', in Becker et al., pp. 242-254.

Bossel, H. (1994), *Modellbildung und Simulation: Konzepte, Verfahren und Modelle zum Verhalten dynamischer Systeme* (Braunschweig, Wiesbaden: Vieweg).

von Böventer, E. and Hampe, J. (1988) *Ökonomische Grundlagen der Stadtplanung; Hannover* (Verlag der Akademie für Raumforschung und Landesplanung).

Bredenkamp, K. and Bredenkamp, J. (1974), 'Was ist Lernen?', in Weinert et al. (eds), pp. 605-630.

von Bredow, R. (2000), 'Genetik "Ist er nicht hübsch?"', in *Der Spiegel* 17: 178-182.

Brewer, W.F. and Treyens, J.C. (1981), 'Role of Schemata in Memory of Places', in *Cognitive Psychology* 13: 207-230.

Brown, R. (1997), *Group Processes. Dynamics within and between Groups* (Oxford: Blackwell).

Bührke, Th. (2000), 'Die verborgenen Dimensionen', in *Bild der Wissenschaft*, 10: 62-63.

Bunge, M. (1974a), *Treatise on Basic Philosophy (Volume 1): Semantics I* (Dordrecht, Bosten: Reidel).

Bunge, M. (1974b), *Treatise on Basic Philosophy (Volume 2): Semantics II* (Dordrecht, Bosten: Reidel).

Bunge, M. (1977), *Treatise on Basic Philosophy (Volume 3): Ontology I* (Dordrecht, Bosten: Reidel).

Bunge, M. (1979), *Treatise on Basic Philosophy (Volume 4); Ontology II. A World of Systems.* (Dordrecht, Bosten: Reidel).

Bunge, M. (1983a), *Epistemologie: aktuelle Fragen der Wissenschaftstheorie.* (Mannheim: Bibliographisches Institut).

Bunge, M. (1983b), *Treatise on Basic Philosophy (Volume 5): Epistemology I: Exploring the World* (Dordrecht, Bosten: Reidel).

Bunge, M. (1987), *Kausalität, Geschichte und Probleme* (Tübingen: Mohr).

Bunge, M. (1989), *Treatise on Basic Philosophy (Volume 8): The Good and the Right* (Dordrecht, Bosten: Reidel).

Bunge, M. (1996), *Finding Philosophy in Social Science*. (New Haven: Yale University Press).

Bunge, M. (1998), *Philosophy of Science, Volume 2: From Explanation to Justification*. (New Brunswick, London: Transaction Publishers).

Bunge, M. (1999a), *Dictionary of Philosophy* (New York: Prometheus Books).

Bunge, M. (1999b), 'Mechanism', in Bunge, M. (1999), *The Sociology-Philosophy Connection* (New Brunswick, London: Transaction Publishers), pp. 17ff.

Bunge, M. and Mahner, M. (2004), *Über die Natur der Dinge. Materialismus und Wissenschaft* (Stuttgart: Hirzel).

Bußmann, H. (1990), *Lexikon der Sprachwissenschaft* (Stuttgart: Kröner).

Campbell, D.T. and Stanley, J.C. (1966), *Experimental and Quasi-experimental Designs for Research* (Chicago: Rand McNally).

Campbell, H. (2003), 'Talking the same words but speaking different languages: the need for more meaningful dialogue', in *Planning Theory and Practice*. 4: 389-392.

Castells, M. (1977), *The Urban Question: A Marxist Approach* (London: Edward Arnold).

Castells, M. (1978), 'The Social Function of Urban Planning: State Action in the Urban-Industrial Development of the French Northern Coastline', in Castells, M. *City, Class and Power* (London: Macmillan), pp. 62-92.

Catton, W.R. (1980), *Overshoot: The Ecological Basis of Revolutionary Change* (Urbana: University of Illinois Press).

Cherniak, Ch. (1992), *Minimal Rationality* (Cambridge, MA: MIT Press).

Clavel, P. (1994), 'The Evolution of Advocacy Planning', in *Journal of the American Planning Association* (60) 2, Spring; pp. 146-149.

D'Abro, A. (1939), *The Decline of Mechanism (in Modern Physics)* (New York: Van Nostrand).

Davidoff, P. (1965), 'Advocacy and Pluralism in Planning', in *Journal of the American Institute of Planners* (31) November, pp. 331-338.

Davidoff, P. and Reiner T.A. (1962), 'A Choice Theory of Planning', in *Journal of the American Institute of Planners* (28), May, pp. 103-115.

Davy, B. (1997), *Essential Injustice* (New York: Springer).

DeBono, E. (1972), *Die 4 richtigen und 5 falschen Denkmethoden* (Reinbek: Rowohlt).

Deutscher, J. (1973), *What we Say/What we Do* (Glenview, Illinois: Scott, Foresman and Company).

DIN 2330 (1993), *Begriffe und Benennungen; Allgemeine Grundsätze* (Berlin: Deutsches Institut für Normung).

Dören, B. (1998), 'Chemnitz – Leitlinien zur Entwicklung einer fragmentierten Stadt', in Becker et al., pp. 188-194.

Dörner, D. (1976), *Problemlösen als Informationsverarbeitung* (Stuttgart: Kohlhammer).

Dörner, D. (1989) *Die Logik des Misslingens, Strategisches Denken in komplexen Situationen* (Reinbek bei Hamburg: Rowohlt).

Dörner, D. (1995), 'Problemlösen und Gedächtnis', in Dörner, D. and van der Meer, E. (Hrsg.) (1995), *Das Gedächtnis* (Göttingen: Hogrefe), pp. 295-320.

Dym, C.L. (1994), *Engineering Design: A Synthesis of Views* (Cambrigde: Cambridge University Press).

Eco, U. (1977), *Zeichen. Einführung in einen Begriff und seine Geschichte* (Frankfurt am Main: Suhrkamp).

Eco, U. (1991), *Einführung in die Semiotik* (München: Fink).

Eekhoff, J., Heidemann, C. and Strassert, G. (1981), *Kritik der Nutzwertanalyse* Karlsruhe: Institut für Regionalwissenschaft, Diskussionspapier Nr. 11.

Eppinger, J. (1998), 'Hannover – Weltausstellung und Stadtzukunft', in Becker et al., pp. 216-226.

Esser, H. (1993), *Soziologie. Allgemeine Grundlagen* (Frankfurt am Main: Campus).

Etzioni, A. (1967), 'Mixed-Scanning: A "Third" Approach To Decision-Making', in *Public Administration Review* (27): 385-392.

Eysenck, M.W. and Keane, M.T. (1998), *Cognitive Psychology* (Hove: Psychology Press).

Faludi, A. (1973), *A Reader in Planning Theory* (New York: Pergamon Press).

Faludi, A. (1986), *Critical Rationalism and Planning Methodology. Research in Planning and Design* (London: Pion).

Faludi, A. (1987), *A Decision-centred View of Environmental Planning* (Oxford: Pergamon Press).

Faludi, A. (1996), 'Rationality, Critical Rationalism, and Planning Doctrine', in Mandelbaum, Mazza and Burchell, pp. 65-82.

Feldtkeller, A. (1998), 'Französisches Viertel Tübingen – "Mischen Sie mit!"', in Becker et al., pp. 270-278.

Feldtkeller, Ch. (1989), *Der architektonische Raum: eine Fiktion* (Braunschweig: Vieweg).

Feyerabend, P. (1975/1979), *Wider den Methodenzwang. Skizze einer anarchistischen Erkenntnistheorie [Against Method. Outline of an Anarchistic Theory of Knowledge]*, (Frankfurt am Main: Suhrkamp).

Fischer, F. and Forester, J. (eds) (1993), *The Argumentative Turn in Policy Analysis and Planning* (Durham: Duke University Press).

Flade, A. (1990), 'Kriminalität und Vandalismus', in Kruse, L. et al. (eds), pp. 518-524.

Flyvbjerg, B. (1998), 'Empowering Civil Society: Habermas, Foucault and the Question of Conflict', in Douglass, M. and Friedmann, J. (eds), *Cities for Citizens. Planning and the Rise of Civil Society in a Global Age* (Chichester, New York: John Wiley & Sons), pp. 187-211.

Fodor, J.A. (1979), *The Modularity of Mind* (Cambridge, MA: MIT Press).

Forester, J. (1989), *Planning in the Face of Power* (Berkeley: University of California Press).

Forester, J. (1993), *Critical Theory, Public Policy, and Planning Practice* (New York: State University of New York Press).

Foucault, M. (1982), 'The Subject and Power', in Dreyfus, H. and Rabinow, P. (eds), *Michel Foucault: Beyond Structuralism and Hermeneutics* (Brighton: Harvester Press), pp. 214-232.

Fredrickson, J.W. and Mitchell, T.R. (1984), 'Strategic decision processes: Comprehensiveness and performance in an industry with an unstable environment', in *Academy of Management Journal*, (27):399-423.

Frege, G. (1952), 'On sense and reference', in Geach, P. and Black, M. (eds), *Translations from the Philosophical Writings of Gottlob Frege* (Oxford: Basil Blackwell).

Friedmann, J. (1973), *Retracking America: A Theory of Transactive Planning* (New York: Anchor Press).

Friedmann, J. (1996), 'Two Centuries of Planning Theory: An Overview', in Mandelbaum, Mazza und Burchell, pp. 10-29.

Friedrich Ebert Stiftung (2000), Fachtagung: Theoretische Grundlagen der Städtebau- und Stadtentwicklungspolitik. Bonn, 23 November 2000.

Fritz-Haendeler, R. (1998), 'Regionale Leitbildentwicklungen in Brandenburg – ein Verständigungsprozess', in Becker et al., pp. 228-238.

Gabriel, G. (2004), 'Postmoderne', in Mittelstraß, J. (ed.), *Enzyklopädie Philosophie und Wissenschaftstheorie (Band 3)*, (Stuttgart, Weimar: Metzler), p. 306.

Gentner, D. and Stevens, A.L. (1983), *Mental Models* (Hillsdale, NJ: Lawrence Erlbaum).

Giddens, A. (1988), *Die Konstitution der Gesellschaft. Grundzüge einer Theorie der Strukturierung* (Frankfurt am Main: Campus).

Glasersfeld, E. von (1997), *Wege des Wissens* (Heidelberg: Carl Auer).

Gollwitzer, P.M. (1996), 'Das Rubikonmodell der Handlungsphasen', in Kuhl. J. and Heckhausen, H. (1996), *Motivation, Volition und Handlung; Enzyklopädie der Psychologie, Themenbereich C, Theorie und Forschung; Ser. 4, Motivation und Emotion, Band 4*, pp. 531-582.

Goodman, R. (1972), *After the Planners* (London: Penguin).

Graumann, C.F. (1975), 'Person und Situation', in Lehr, U. and Weinert, F.E, *Entwicklung und Persönlichkeit* (Stuttgart: Kohlhammer), pp. 15-24.

Groeben, N. and Westmeyer, H. (1975), *Kriterien psychologischer Forschung* (München: Juventa).

Habermas, J. (1981), *Theorie des kommunikativen Handelns, I-II* (Frankfurt: Suhrkamp).

Habermas, J. (1983), *Moralbewusstsein und kommunikatives Handeln* (Frankfurt: Suhrkamp).

Hagen, T. (1988), *Wege und Irrwege der Entwicklungshilfe* (Zürich: Verlag Neue Zürcher Zeitung).

Halentz, R. (1997), 'Haben Tiere doch ein Bewusstsein?', in *Bild der Wissenschaft* (7): 60-63.

Hall, P. (1988), *Cities of Tomorrow. An Intellectual History of Urban Planning and Design in the Twentieth Century* (Oxford: Basil Blackwell).

Harper, Th. L. and Stein, St. M. (2006), *Dialogical Planning in a Fragmented Society* (New Brunswick: Center for Urban Policy Research Press).

Hayek, F.A. (1976), *Law, Legislation and Liberty, Vol. 2: The Mirage of Social Justice* (Chicago: University of Chicago Press).

Hayes, N. (1996), *Foundations of Psychology* (Walton-on-Thames: Nelson).

Healey, P. (1997), *Collaborative Planning: Shaping Places in Fragmented Societies* (Basingstoke: Macmillan).

Heckhausen, H. (1974), 'Anlage und Umwelt als Ursache für Intelligenzunterschiede', in Weinert et al., pp. 275-312.

Hedström, P. and Swedberg, R. (eds) (1998), *Social Mechanisms* (New York: Cambridge University Press).

Heidemann, C. (1985), 'Zukunftswissen und Zukunftsgestaltung – Planung als verständiger Umgang mit Mutmaßungen und Gerüchten', in Daimler-Benz-Aktiengesellschaft (Hrsg.), *Langfristprognosen: Zahlenspielerei oder Hilfsmittel für die Planung?* (Düsseldorf: VDI-Verlag), pp. 47-62.

Heidemann, C. (1990), Darstellung, Verständnis und Verständigung. Hinweise zum Umgang mit semiotischen Tücken in der Planung. (Karlsruhe: Institut für Regionalwissenschaft), Diskussionspapier Nr. 18.

Heidemann, C. (1992), Regional Planning Methodology. The First & Only Annotated Picture Primer on Regional Planning. (Karlsruhe: Institut für Regionalwissenschaft), Discussion Paper Nr. 16 .

Heidemann, C. (1993), Die Entwicklungsvokabel – Redenschmuck oder Gedankenstütze?; Institut für Regionalwissenschaft der Universität Karlsruhe (Hrsg.); Diskussionspapier Nr. 23.

Heidemann, C. (1995), *Vorlesung Planungstheorie* (Karlsruhe: Institut für Regionalwissenschaft).

Hellbrück, H. and Fischer, M. (1999), *Umweltpsychologie* (Göttingen: Hogrefe).

Hellweg, U. (1998), 'Stadtumbau auf historischem Grundriß – die neue Unterneustadt in Kassel', in Becker et al. (1998), pp. 280-286.

Heskin, A. (1980), 'Crisis and response: An Historical Perspective on Advocacy Planning', in *Journal of the American Planning Association*, (46)1: 50-63.

Hoffmann, J. (1986), *Die Welt der Begriffe* (Berlin: VEB Deutscher Verlag der Wissenschaften).

Huber, R. (1976), 'Einführung in die Systemtechnik Grundlagen, Möglichkeiten und Grenzen', in VDI (Hrsg.), *Grundlagen und Anwendungen der Systemtechnik als rationales Hilfsmittel für Wirtschaft, Staat und Forschung* (Düsseldorf: VDI-Verlag), pp. 5-17.

Hudson, B. M. (1979), 'Comparison of Current Planning Theories: Counterparts and Contradictions', in *Journal of the American Planning Association* (45): 387-405.

Innes, J. E. (1995), 'Planning Theory's Emerging Paradigm: Communicative Action and Interactive Practice', in *Journal of Planning Education and Research* pp. 183-189.

Ipsen, D. (1998), 'Moderne Stadt – was nun', in Becker et al., pp. 42-54.

Janis, I.L. (1972), *Victims of Groupthink* (Boston: Houghton Mifflin).

Jantsch, E. (1992), 'System, Systemtheorie', in Seiffert und Radnitzky, pp. 329-338.

Jennings, D.L., Amabile, T.A. and Ross, L. (1982), 'Informal covariation assessment: Data-based versus theory-based judgements', in Kahneman, D., Slovic, P. and Tversky, A. (eds) (1982), *Judgement under uncertainty: Heuristics and biases.* (Cambridge: Cambridge University Press) pp. 211-230.

Jessen, J. (1996), 'Der Weg zur Stadt der kurzen Wege – versperrt oder nur lang? Zur Attraktivität eines Leitbildes', in *Archiv für Kommunalwissenschaften* (AfK) (35)I: 1-19.

Johnson-Laird, P.N. (1983), *Mental Models* (Cambridge, MA: Harvard University Press).

Kambartel, F. (1996), 'Kritische Theorie', in Mittelstraß (Band 4), pp. 270-271.

Kant, E. (1787/1963). *Critique of Pure Reason* (London: Macmillan).

Klix, F. (1992). *Die Natur des Verstandes* (Göttingen: Hogrefe).

Kollhoff, H. (1998), 'Beitrag zum Streitgespräch "Zukunft der Stadt – leitbildorientiert oder nicht?" zwischen Manfred Birk, Hans Kollhoff, Willi Polte, Christiane Thalgott, Rudolf Schäfer u.a', in Becker et al., pp. 81-108.

Koppenjan, J. and Klijn, E.-H. (2004), *Managing Uncertainties in Networks. A Network approach to Problem Solving and Decision Making* (London: Routledge).

Koschitz, P. (1993), *Zur Darstellung raumplanerischer Problemsituationen. ORL-Bericht 90* (Zürich: Verlag der Fachvereine).

Krätke, St. (1995), *Stadt – Raum – Ökonomie; Einführung in aktuelle Problemfelder der Stadtökonomie und Wirtschaftsgeographie.* (Basel, Boston: Birkhäuser).

Krumholz, N. (1994), *Dilemmas of Equity Planning: A Personal Memoir'*, in *Planning Theory* (10/11): 45-58.

Krumholz, N. and Forester, J. (1990), *Making Equity Planning Work* (Philadelphia: Temple University Press).

Kruse, L., Graumann, C.-F. and Lantermann, E.-D. (Hrsg.) (1990), *Ökologische Psychologie. Ein Handbuch in Schlüsselbegriffen* (München-Weinheim: Psychologie Verlags Union).

Kuhn, Th. S. (1962/1981), *Die Struktur wissenschaftlicher Revolutionen, The Structure of Scientific Revolutions* (Frankfurt am Main: Suhrkamp).

Kunst, F. (1998), Leitbilder für Berliner Stadträume – der "innovative Nordosten" und die "Wissenschaftsstadt Adlershof", in Becker et al., pp. 206-214.

Lefebvre, H. (1968), *Le Droit à la ville* (Paris: Éditions Anthropos).

Lefebvre, H. (1972), *Espace et politique: Le Droit à la ville II* (Paris: Éditions Anthropos).

Lendi, M. (1998), 'Grundkonstanten der Raumplanung. Bauen und Umnutzung von Bauten außerhalb des Siedlungsgebietes – Fragwürdigkeiten', in DISP (34) 132: pp. 25-34.

Lenk, H. and Spinner, H.F. (1989), 'Rationalitätstypen, Rationalitätskonzepte und Rationalitätstheorien im Überblick; Zur Rationalismuskritik und Neufassung der "Vernunft heute", in Stachowiak, H. (ed.), pp. 1-31.

Lindblom, C. (1959), 'The Science of "Muddling Through"', in Stein, J.M. (ed.) (1995), *Classic Readings in Urban Planning* (New York: McGraw-Hill), pp. 35-48.

Lorenz, K. (1996), 'Tautologie', in Mittelstraß, pp. 213-214.

Luhmann, N. (1979), 'Öffentliche Meinung', in: Langenbucher, W.R. (ed.), *Politik und Kommunikation. Über die öffentliche Meinungsbildung* (München, Zürich: Piper).

Luhmann, N. (1996), *Soziale Systeme* (Frankfurt am Main: Suhrkamp Taschenbuch Wissenschaft).

Mahner, M. and Bunge, M. (1997), *Foundations of Biophilosophy* (Berlin, Heidelberg: Springer).

Malik, F. (1999), *Management-Perspektiven* (Bern, Stuttgart: Paul Haupt).

Mandelbaum, S.J. (1979), 'A complete general theory of planning is impossible', in *Policy Sciences* (11)1 (August): 59-71.

Mandelbaum, S.J., Mazza, L. and Burchell, R.W. (eds) (1996), *Explorations in Planning Theory* (New Brunswick, New Jersey: Rutgers, Center for Urban Policy Research).

March, J.G. (1978), 'Bounded rationality, ambiguity, and the engineering of choice', in *The Bell Journal of Economics* (9): 587-608.

March, J.G. (1982), 'Theories of Choice and Making Decisions', in *Society* (1)20: 29-39.

Maurer, J. (1993), 'Über die Methodik der Raumplanung', in Strohschneider und von der Weth, pp. 208-218.

Maurer, J. (1998), 'Strategische und organisatorische Anforderungen zur Konkretisierung und Umsetzung von Leitbildern', in Becker et al., pp. 72-80.

Mayntz, R. (1976), 'Conceptual models of organisational decision-making and their application to the policy process', in Hofstede, G. and Kassem, M.S. (eds), *European Contributions to Organization Theory* (Assen: Van Gorcum), pp. 114-125.

McCloskey, M. (1983), 'Intuitive physics', in *Scientific American*, (24):122-130.

Meyerson, M. and Banfield, E.C. (1955), *Politics, Planning, and the Public Interest: The Case of Public Housing in Chicago* (London: The Free Press of Glencoe).

Mill, J.S. (1859)/1977), 'On Liberty', in *Collected Works, Vol. XVIII* (Toronto: University of Toronto Press).

Miller, G.A. (1981), *Language and Speech* (San Francisco: Freeman).

Miller, G.A., Galanter, E. and Pribam, K.K. (1960/1991), *Strategien des Handelns, Pläne und Strukturen des Verhaltens*. 2nd Edition [Plans and Structure of Behavior], (Stuttgart: Klett-Cotta).

Min, L. (2001), *Certainty as a Social Metaphor: The Social and Historical Production of Certainty in China and the West* (Westport: Greenwood).

Minsky, M. (1975), 'A framework for representing knowledge', in Winston, P.H. (ed.) (1975), *The Psychology of Computer Vision* (New York: McGraw-Hill).

Mittelstraß, J. (1996) (ed.), *Enzyklopädie Philosophie und Wissenschaftstheorie (Band 4)*, (Stuttgart, Weimar: Metzler).

Möller, K.P. (1999), 'Ist Nachhaltigkeit nur eine Worthülse?', in *Bild der Wissenschaft* Nr.1, p. 12.

Mönninger, M. (1999) (ed.), *Stadtgesellschaft* (Frankfurt am Main: Suhrkamp).

Muller, J. (1992), 'From survey to strategy: twentieth century developments in western planning method', in *Planning Perspectives. An International Journal of History, Planning and the Environment* (7):125-155.

Myers, D. and Kitsuse, A. (2000), 'Constructing the Future in Planning: A Survey of Theories and Tools', in *Journal of Planning Education and Research* (19)3: 221-231.

Neisser, U. (1979), *Kognition und Wirklichkeit* (Stuttgart: Klett-Cotta).

Newell, A. and Simon, H.A. (1972), *Human Problem Solving* (Englewood Cliffs: Prentice-Hall).

Nisbett, R.E. and DeCamp Wilson, T. (1977), 'Telling More Than We can Know: Verbal Reports on Mental Processes', in *Psychological Review*, (84)3: 231-259.

Nozick, R. (1974), *Anarchy, State and Utopia* (New York: Basic Books).

Ortony, A. (1975), 'Why metaphors are necessary and not just nice', in *Educational Theory* (25)1: 45-53.

Parsons, T. (1949), *The Structure of Social Action* (Glencoe, Illinois: The Free Press).

Patton, C.V. and Sawicki, D.S. (1993), *Basic Methods of Policy Analysis and Planning* (Englewood Cliffs: Prentice Hall).

Paulus, J. (1994), 'Auf Gedanken-Fang', in *Spektrum der Wissenschaft* 12: 112.

Peattie, L. (1968), Reflections on Advocacy Planning', in *Journal of the American Institute of Planning* (31)4: 331-338.

Piaget, J. (1967), *The Child's Conception of the World* (Totowa, NJ: Littlefield, Adams).

Piaget, J. (1970), 'Piaget's theory', in Mussen, J. (ed.), *Carmichael's Manual of Child Psychology* (Vol.1), (New York: Basic Books).

Piaget, J. (1974), *Biologie und Erkenntnis. Über die Beziehungen zwischen organischen Regulationen und kognitiven Prozessen* (Frankfurt: Fischer).

Piaget, J. (1976), *Die Äquilibration der kognitiven Strukturen* (Stuttgart: Klett).

Popper, K. (1987), *Das Elend des Historizismus* (Tübingen: Mohr).

Poulton, M.C. (1991a), 'The case for a positive theory of planning. Part 1: What is wrong with planning theory?', in *Environment and Planning B: Planning and Design*, (18): 225-232.

Poulton, M.C. (1991b), 'The case for a positive theory of planning. Part 2: A positive theory of planning', in *Environment and Planning B: Planning and Design*, (18): 263-275.

Pylyshyn, Z.W. (1986), *Computation and Cognition (Toward a Foundation for Cognitive Science)*, (London: The MIT Press, Bradford Books).

Raith, E. (2000), *Stadtmorphologie* (Wien, New York: Springer).

Reason, J. (1990), *Human Error* (Cambridge: Cambridge University Press).

Reason, J. (1994), *Menschliches Versagen: psychologische Risikofaktoren und moderne Technologien* (Heidelberg: Spektrum Akademischer Verlag).

Reck, H. (1994), Umweltverträglichkeitsuntersuchung und Landschaftspflegerischer Begleitplan im Straßenbau: Entwicklung eines Handlungsrahmens zur Ermittlung und Beurteilung straßenbedingter Auswirkungen auf Pflanzen, Tiere und ihre Lebensräume; Stuttgart: Dissertation an der Fakultät für Architektur und Stadtplanung.

Rescher, N. (1998), *Complexity* (New Brunswick, New Jersey: Transaction Publishers).

Richardson, H.W. (1978), '"Basic" Economic Activities in Metropolis', in Leven, C.L. (1978) (ed.), *The Mature Metropolis* (Lexington, Toronto).

Riedl, R. (1980), *Biologie der Erkenntnis. Die stammesgeschichtlichen Grundlagen der Vernunft* (Berlin, Hamburg: Paul Parey).

Rittel, H. (1970), 'Der Planungsprozess als iterativer Vorgang von Varietätserzeugung und Varietätseinschränkung', in Institut für Grundlagen der Modernen Architektur (Hrsg.), *Arbeitsberichte zur Planungsmethodik* (4): 17-31.

Rittel, H. (1972), 'On the Planning Crisis: Systems Analysis of the First and Second Generations', in Bedriftsøkonomen, No.8, October; pp. 390-396.

Rittel, H. and Webber, M. (1973), 'Dilemmas in a General Theory of Planning', in *Policy Sciences* 4(2): 155-169.

Rodi, F. (1992), 'Semiotik', in Seiffert und Radnitzky, pp. 297-301.

Ros, A. (1989), *Begründung und Begriff: Wandlungen des Verständnisses begrifflicher Argumentation; Band 1: Antike. Spätantike und Mittelalter* (Hamburg: Felix Meiner).

Ros, A. (1990a), *Begründung und Begriff: Wandlungen des Verständnisses begrifflicher Argumentation; Band 2: Neuzeit* (Hamburg: Felix Meiner).

Ros, A. (1990b), *Begründung und Begriff: Wandlungen des Verständnisses begrifflicher Argumentation; Band 3: Moderne* (Hamburg: Felix Meiner).

Rudolph, M. (2000), *Eine kritische Betrachtung der nachhaltigen (Stadt)Entwicklung.* (Stuttgart: Diplomarbeit am Institut für Grundlagen der Planung der Universität Stuttgart).

Rumelhart, D.E. (1975), 'Notes on a schema for stories', in Bobrow, D.G. and Collins, A. (eds), *Representation and Understanding: Studies in Cognitive Science* (New York: Academic Press).

Rumelhart, D.E. (1980), 'Schemata: The basic building blocks of cognition', in Spiro, R., Bruce, B. and Brewer, W. (eds), *Theoretical Issues in Reading Comprehension* (Hillsdale, NJ: Lawrence Erlbaum).

Rumelhart, D.E. and Ortony, A. (1977), 'The representation of knowledge in memory', in Anderson, R.C., Spiro, R.J. and Montague, W.E. (eds), *Schooling and the Acquisition of Knowledge* (Hillsdale, NJ: Lawrence Erlbaum).

Saegert, S. (1985), 'The role of housing in the experience of dwelling', in Altman, I. and Carol, M. W. (eds), *Home Environments* (New York: Plenum Press).

Saltzwedel, J. (1998), 'Karneval der Ideen',in *Der Spiegel* Nr. 11 (1998), pp. 210-213.

Sandercock, L. (1998), 'The Death of Modernist Planning: Radical Praxis for a Postmodern Age', in Douglass, M. and Friedmann, J. (eds), *Cities for Citizens: Planning and the Rise of Civil Society in a Global Age* (Chichester, New York: John Wiley & Sons), pp. 163-184.

Sandercock, L. (1998a), *Towards Cosmopolis: Planning For Multicultural Cities* (Chichester: John Wiley & Sons).

Sauga, M. (2000), 'Riesters Rententrick', in *Der Spiegel*, (12) 28.

Schäfer, M. and Schön, S. (2000), *Nachhaltigkeit als Projekt der Moderne* (Berlin: Edition Sigma).

Schank, R.C. (1972), 'Conceptual dependency: A theory of natural language understanding', in *Cognitive Psychology* (3): 552-631.

Schank, R.C. and Abelson, R.P. (1977), *Scripts, Plans, Goals and Understanding* (Hillsdale, NJ: Lawrence Erlbaum).

Schelling, T.C. (1978), *Micromotives and Macrobehavior* (New York: Norton).

Schimank, U. (1996), *Theorien gesellschaftlicher Differenzierung* (Opladen: Leske und Budrich).

Schlicksupp, H. (1992), *Ideenfindung* (Würzburg: Vogel).

Schneider, B. (1998), 'Städtebauliche Leitbilder – Weltbilder, Trugbilder, Selbstbildnisse', in Becker et al., pp. 124-134.

Schön, D.A. and Rein, M. (1994), *Frame Reflection. Toward the Resolution of Intractable Policy Controversies* (New York: Basic Books).

Schönwandt, W.L. (1982), Hinweise der Sozialwissenschaften zur Wohnungsplanung; Bad Godesberg: Schriftenreihe "Bau- und Wohnforschung" des Bundesministers für Raumordnung. Bauwesen und Städtebau, 04.077.

Schönwandt, W.L. (1986), *Denkfallen beim Planen* (Braunschweig: Vieweg).

Schönwandt, W.L. and Wasel, P. (1997), 'Das semiotische Dreieck – ein gedankliches Werkzeug beim Planen, Teil I', in Bauwelt (88), Heft (19):1028-1042; Teil II, in Bauwelt (1997) (88), Heft (20): 1118-1130.

Schönwandt, W.L. (1999), 'Grundriss einer Planungstheorie der "dritten Generation"', in DISP 136/137, April (1999) (35. Jahrgang), pp. 25-35.

Schönwandt, W.L. (2000), 'Sieben Planungsmodelle', in *RaumPlanung* 93, December, pp. 292-299.

Schönwandt, W. and Jung, W. (2006), 'The Turn to Content', in Selle, K. (ed.), *Planung neu denken? Band 1; Zur räumlichen Entwicklung beitragen. Konzepte – Theorien – Impulse* (Dortmund: Verlag Dorothea Rohn).

Schroeder-Heister, P. (1984), 'Kausalanalyse', in Mittelstraß, J., *Enzyklopädie Philosophie und Wissenschaftstheorie, Band 2* (Mannheim: Bibliographisches Institut), pp. 371-372.

Seiffert, H. and Radnitzky, G. (1992) (eds), *Handlexikon zur Wissenschaftstheorie* (München: Deutscher Taschenbuch Verlag).

Selle, K. (1994), *Expositionen. Materialen zur Diskussion um die Expo 2000* (Dortmund: Vertrieb für Bau- und Planungsliteratur).

Selle, K. (1997), 'Planung und Kommunikation', in *DISP* 129 (33): 40-47.

Seni, D.A. (1996), 'Planning Theory or the Theory of Plans?', in Kuklinski, A. *Production of Knowledge and the Dignity of Science* (Warsaw: Rewasz), pp. 147-159.

Siegwart, G. (1996), 'Systemtheorie', in Mittelstraß, pp. 190-194.

Signer, R. (1994), Argumentieren in der Raumplanung; Zürich: Dissertation an der Eidgenössischen Technischen Hochschule Zürich.

Simon, H.A. (1947), *Administrative Behavior* (New York: Macmillan).

Simon, H.A. (1965), *The Shape of Automation for Men and Management* (New York: Harper & Row).

Simon, H.A. (1968), *The Science of the Artificial* (Cambridge, Massachusetts: MIT Press).

Simon, H.A. (1973), 'The Structure of Ill-structured Problems', in *Artificial Intelligence* (4): 181-201.

Simon, H.A. (1976), *Administrative Behavior* (New York: Free Press).

Sokal, A.; Bricmont, J. (1998), *Fashionable Nonsense* (New York: Picador).

Sommer, H. (1998), *Projektmanagement im Hochbau* (Berlin: Springer).

Sommer, R. (1969), *Personal Space – The Behavioral Basis of Design* (Englewood Cliffs: Prentice Hall).

Sorensen, A.D. (1983), 'Toward a Market Theory of Planning', in *The Planner* 69(3): 78-80.

Sorensen, A.D. and Day, R.A. (1981), 'Libertarian Planning', in *Town Planning Review* (52): 390-402.

Spaemann, R. (1992), 'Teleologie', in Seiffert, H. and Radnitzky, G. *Handlexikon der Wissenschaftstheorie* (München: Deutscher Taschenbuch Verlag), pp. 366-368.

Stachowiak, H. (1989) (ed.), *Pragmatik, Handbuch pragmatischen Denkens, Band III Allgemeine philosophische Pragmatik* (Hamburg: Felix Meiner).

Stachowiak, H. (1992), 'Planung', in Seiffert and Radnitzky, pp. 262-267.

Stein, N.L. and Glenn, C.G. (1979), 'An analysis of story comprehension in elementary school children', in Freedle, R. (ed.), *Multidisciplinary Perspectives in Discourse Comprehension* (New Jersey: Ablex).

Stegmüller, W. (1983), *Probleme und Resultate der Wissenschaftstheorie und Analytischen Philosophie. Band I: Erklärung – Begründung – Kausalität. Studienausgabe, Teil A* (Berlin: Springer).

Stone, D.A. (1988), *Policy Paradox and Political Reason* (Glenview, Illinois: Scott Foresman).

Strassert, G. (1995), *Das Abwägungsproblem bei multikriteriellen Entscheidungsproblemen: Grundlagen und Lösungsansatz unter besonderer Berücksichtigung der Regionalplanung* (Frankfurt am Main: Peter Lang).

Strohmeyer, K. (1999), 'Kursbücher für das Chaos der Städte', in *Die Zeit*, Nr. 41, 7 October, p. 60.

Strohschneider, St. and von der Weth, R. (1993) (eds.), *Ja, mach nur einen Plan. Pannen und Fehlschläge – Ursachen, Beispiele, Lösungen* (Bern: Hans Huber).

Süskind, P. (1985), *Das Parfüm. Die Geschichte eines Mörders* (Zürich: Diogenes).

Tank, H. (1987), *Stadtentwicklung – Raumnutzung – Stadterneuerung* (Göttingen: Vandenhoeck & Ruprecht).

Tewdwr-Jones, M. and Allmendinger, Ph. (2002), 'Communicative Planning, Collaborative Planning and the Post-Positivist Planning Theory Landscape', in Allmendinger and Tewdwr-Jones, pp. 206-216.

Thiel, Ch. (1972), *Grundlagenkrise und Grundlagenstreit* (Meisenheim am Glan: Hain).

Thomas, M.J. (1982), 'The Procedural Planning Theory of A. Faludi', in Paris, Ch. (ed.), *Critical Readings in Planning Theory* (Oxford: Pergamon Press), pp. 13-25.

Thorndyke, P.W. (1977), 'Cognitive structures in comprehension and memory of narrative discourse', in *Cognitive Psychology* (9): 77-110.

Toulmin, St. (1972/1978), *Kritik der kollektiven Vernunft* [Human Understanding, Volume I, General Introduction and Part I: The Collective Use and Evolution of Concepts], (Frankfurt am Main: Suhrkamp).

Uexküll, J. von (1928/1973), *Theoretische Biologie* (Frankfurt am Main: Suhrkamp).

UVF (Umlandverband Frankfurt) (1984), *Flächennutzungsplan Erläuterungsbericht* (Frankfurt am Main: Selbstverlag).

Venturi, M. (1998), 'Leitbilder? Für welche Städte?', in Becker et al., pp. 56-70.

Voigt, A. and Walchhofer, H.P. (eds) (2000), 'Planungstheorie – Bebauungsplanung – Projektsteuerung', in Schriftenreihe des Instituts für örtliche Raumplanung (IFOER), Technische Universität Wien, E268-3.

Vollmer, G. (1988), *Was können wir wissen? Band 1. Die Natur der Erkenntnis* (Stuttgart: Hirzel).

Vollmer, G. (1993), *Wissenschaftstheorie im Einsatz* (Stuttgart: Hirzel).

Warburton, N. (1996), *Thinking from A to Z* (London: Routledge).

Weaver, C., Jessop, J. and Das, V. (1985), 'Rationality in the public interest: notes towards a new synthesis', in Breheny, M. and Hooper, A. (eds), *Rationality in Planning* (London: Pion).

Wegener, M. (1994), 'Operational Urban Models: State of the Art', in *Journal of the American Planning Association*, Winter (60)1: 17-29.

Weik, K.E. (1985), *Der Prozeß des Organisierens* (Frankfurt am Main: Suhrkamp).

Weinert, F.E.,Graumann, C.F., Heckkausen, H. and Hofer, M. (eds), *Pädagogische Psychologie, Band 1 und 2* (Frankfurt am Main: Fischer).

von der Weth, R. (1999), 'Design instinct? – the development of individual strategies', in *Design Studies* 20(5): 453-463.

Wiener, N. (1948/1968), *Cybernetics* (New York: Wiley).

Wildavsky, A. (1979), *Speaking Truth to Power* (Boston: Little Brown).

Winkelmann, U. (1998), *Modelle als Instrument der räumlichen Planung; in: Akademie für Raumforschung und Landesplanung* (ARL) (Hrsg.) (1998), *Methoden und Instrumente räumlicher Planung* (Hannover: Verlag der ARL), pp. 51-66.

Winograd, T. and Flores F. (1989), *Erkenntnis Maschinen Verstehen* (Berlin: Rotbuch).

Wittmann, P. (2000), 'Vorstädte unschuldig an Blechlawine', in *Bild der Wissenschaft* 8/200; pp. 106-107.

Yiftachel, O. (1989), 'Towards a new typology of planning theories', in *Environment and Planning B: Planning and Design* (16)1 January: 23-39.

Zahn, V. (1998), 'Leitbilder für Lübeck – Entwicklungsperspektiven für ein Weltkulturerbe', in Becker et al., pp. 168-186.

Zimbardo, Ph. G. (1992), *Psychologie* (Berlin, Heidelberg, New York: Springer).

Zoche, P. (2000), *Auswirkungen neuer Medien auf die Raumstruktur* (Karlsruhe: Fraunhofer Institut für Systemtechnik und Innovationsforschung).

Zumkeller, D. (2000), 'Verkehr und Telekommunikation. Erste empirische Ansätze und Erkenntnisse', in Jessen, J., Lenz, B. and Voigt, W. (eds), *Neue Medien, Raum und Verkehr. Wissenschaftliche Analysen und praktische Erfahrungen* (Opladen: Leske und Budrich).

Zwicky, F. (1966), *Entdecken, Erfinden, Forschen im morphologischen Weltbild* (München: Knauer).

Index